OCEANOGRAPHY AND OCEAN ENGINEERING

PRIORITIES FOR U.S. OCEAN SCIENCE AND RESEARCH

OCEANOGRAPHY AND OCEAN ENGINEERING

Additional books in this series can be found on Nova's website
under the Series tab.

Additional E-books in this series can be found on Nova's website
under the E-books tab.

OCEANOGRAPHY AND OCEAN ENGINEERING

PRIORITIES FOR U.S. OCEAN SCIENCE AND RESEARCH

RYAN L. MURPHY
EDITOR

Nova Science Publishers, Inc.
New York

LIBRARY OF CONGRESS CATALOGING-IN-PUBLICATION DATA

Priorities for U.S. ocean science and research / editors, Ryan L. Murphy.
p. cm.
Includes index.
ISBN 978-1-61209-706-0 (hardcover)
1. Mariculture--Economic aspects--United States. 2.
Mariculture--Environmental aspects--United States. 3.
Mariculture--Government policy--United States. I. Murphy, Ryan L.
SH34.P755 2011
639.8--dc22

2011003555

Published by Nova Science Publishers, Inc. ✝ New York

CONTENTS

Preface **vii**

Chapter 1 Charting the Course for Ocean Science in the United States for the Next Decade **1**
NSTC Joint Subcommittee on Ocean Science and Technology

Chapter 2 Interagency Oceans and Human Health Research Implementation Plan: A Prescription for the Future **67**
The Interagency Working Group on Harmful Algal Blooms, Hypoxia, and Human Health

Chapter Sources **205**

Index **207**

PREFACE

Understanding society's impact on the ocean and the ocean's impact on society forms the basis for ensuring a clean, healthy and stable ocean environment that can be responsibly used and enjoyed for generations to come. This new book focuses on ocean forecasting, scientific support for ecosystem-based management and ocean-observing capabilities, along with a comprehensive community effort to define ocean research priorities for the U.S. in the next decade.

Chapter 1- Proper stewardship of the ocean is critical to the long-term vitality of the United States. The ocean, consisting of the open ocean, coasts, coastal watersheds, and Great Lakes, provides food and recreation, contributes to the nation's economic engine, is an element of national security, and is a major influence on the global climate system. Despite its vast extent, the ocean is finite and cannot indefinitely absorb all the stresses being placed on it; thus, the United States must commit to protecting the ocean through responsible stewardship and sensible management. Understanding society's impact on the ocean and the ocean's impact on society forms the basis for ensuring a clean, healthy, and stable ocean environment that can be responsibly used and enjoyed for generations to come. The goal of this document is to provide the guidance to build the scientific foundation to improve society's stewardship and use of, and interaction with, the ocean.

Chapter 2- The ocean[1] provides numerous benefits to humankind, including seafood[2], pharmaceuticals and other natural products, recreational opportunities and aesthetic value, shoreline protection, waste assimilation, nutrient cycling, oxygen production, and drinking water[3]. The ocean also moderates climate and contributes substantially to the U.S. economy. However, the ocean poses risks when people are exposed to disease-causing organisms transmitted through or originating in the marine environment, toxins from harmful algal blooms (HABs) and other microorganisms, chemical contaminants, and when catastrophic events such as hurricanes and tsunamis occur.

In: Priorities for U.S. Ocean Science and Research
Editor: Ryan L. Murphy

ISBN: 978-1-61209-706-0

Chapter 1

CHARTING THE COURSE FOR OCEAN SCIENCE IN THE UNITED STATES FOR THE NEXT DECADE

NSTC Joint Subcommittee on Ocean Science and Technology

EXECUTIVE SUMMARY

Proper stewardship of the ocean is critical to the long-term vitality of the United States. The ocean, consisting of the open ocean, coasts, coastal watersheds, and Great Lakes, provides food and recreation, contributes to the nation's economic engine, is an element of national security, and is a major influence on the global climate system. Despite its vast extent, the ocean is finite and cannot indefinitely absorb all the stresses being placed on it; thus, the United States must commit to protecting the ocean through responsible stewardship and sensible management. Understanding society's impact on the ocean and the ocean's impact on society forms the basis for ensuring a clean, healthy, and stable ocean environment that can be responsibly used and enjoyed for generations to come. The goal of this document is to provide the guidance to build the scientific foundation to improve society's stewardship and use of, and interaction with, the ocean.

Three central elements of science and technology will provide the United States with the knowledge and means to redefine its relationship with the ocean for the better. The *capability to forecast key ocean and ocean-influenced processes and phenomena* will change how society takes action in the future, much like weather forecasts do today. Developing the understanding and capability to forecast key processes related to hurricane and severe-storm formation, ocean currents, fish stocks, and human-health hazards, to name a few, will have significant economic, societal, and environmental benefits. Providing the *scientific support for ecosystem-based management* will allow resources to be managed in ways that account for the complex interactions between those resources and other parts of the marine environment, including humans. *Deploying an ocean-observing system* that can accurately describe the state of the ocean will revolutionize the access to and view of the ocean and increase the pace, efficiency, and scope of ocean research. In turn, this observing capability will enable ocean forecasting and ecosystem-based management.

The focus on ocean forecasting, scientific support for ecosystem-based management, and ocean-observing capabilities was born out of a comprehensive community effort to define ocean research priorities for the United States for the next decade. Aspects of these three central elements are evident throughout the twenty (20) national ocean research priorities, oriented around the most compelling issues in key areas of interaction between society and the ocean (societal themes). These priorities, each with equal weight, focus on understanding critical ocean processes and interactions, and applying that understanding toward stewardship and responsible use of the ocean.

Theme 1: Stewardship of Natural and Cultural Ocean Resources

Research Priority 1: Understand the status and trends of resource abundance and distribution through more accurate, timely and synoptic assessments

Research Priority 2: Understand interspecies and habitat/species relationships to support forecasting resource stability and sustainability

Research Priority 3: Understand human-use patterns that may influence resource stability and sustainability

Research Priority 4: Apply advanced understanding and technologies to enhance the benefits of various natural resources from the open ocean, coasts, and Great Lakes

Theme 2: Increasing Resilience to Natural Hazards

Research Priority 5: Understand how hazard events initiate and evolve and apply thatunderstanding to improve forecasts of future hazard events

Research Priority 6: Understand the response of coastal and marine systems to natural hazards and apply that understanding to assessments of future vulnerability to natural hazards

Research Priority 7: Apply understanding to develop multi-hazard risk assessments and support development of models, policies, and strategies for hazard mitigation

Theme 3: Enabling Marine Operations

Research Priority 8: Understand the interactions between marine operations and the environment

Research Priority 9: Apply understanding of environmental factors affecting marine operations to characterize and predict conditions in the maritime domain

Research Priority 10: Apply understanding of environmental impacts and marine operations to enhance the marine transportation system

Theme 4: The Ocean's Role in Climate

Research Priority 11: Understand ocean-climate interactions within and across regions

Research Priority 12: Understand the impact of climate variability and change on the biogeochemistry of the ocean and implications for its ecosystems

Research Priority 13: Apply understanding of the ocean to help project future climate changes and their impacts

Theme 5: Improving Ecosystem Health

Research Priority 14: Understand and predict the impact of natural and anthropogenic processes on ecosystems

Research Priority 15: Apply understanding of natural and anthropogenic processes to develop socioeconomic assessments and models to evaluate the impact of multiple human uses on ecosystems

Research Priority 16: Apply understanding of marine ecosystems to develop appropriate indicators and metrics for sustainable use and effective management

Theme 6: Enhancing Human Health

Research Priority 17: Understand sources and processes contributing to ocean-related risks to human health

Research Priority 18: Understand human health risks associated with the ocean and the potential benefits of ocean resources to human health

Research Priority 19: Understand how human use and valuation of ocean resources can be affected by ocean-borne human health threats and how human activities can influence these threats

Research Priority 20: Apply understanding of ocean ecosystems and biodiversity to develop products and biological models to enhance human well-being

Four near-term priorities, reflecting efforts to be pursued over the next two to five years, were developed to initiate rapid progress towards the 20 ocean research priorities.

1. Forecasting the Response of Coastal Ecosystems to Persistent Forcing and Extreme Events

Understanding and forecasting the response of natural and constructed landscapes and ecosystems to extreme weather events, natural disasters, and changing ocean conditions will inform hazard mitigation and response plans, support navigation safety, and assist regional resource managers and public health officials in sustaining ecosystem and public health and promoting hazard resilience.

2. Comparative Analysis of Marine Ecosystem Organization

Marine ecosystem management can be improved by elucidating their underlying dynamics at a variety of scales through development of dynamics models and their application in comparisons of managed ecosystems. This effort will provide a greater basic understanding of ecosystem processes and practical tools for evaluating the effectiveness of local and regional ecosystem-based management efforts.

3. Sensors for Marine Ecosystems

New sensor capabilities are necessary to realize the full potential of *in situ* ocean observing networks and satellite-based observations. These capabilities will enable expanded and more accurate information on the ocean environment to better define management options, understand processes that influence ecosystem productivity, serve as the basis for forecasting ocean-related risks to human health and safety, and shed light on the impact of climate variability and change on the ocean, marine life, and humans.

4. Assessing Meridional Overturning Circulation Variability: Implications for Rapid Climate Change

The meridional overturning circulation (MOC) of the Atlantic Ocean, an element of the global-scale ocean circulation responsible for long-term climate variations, has been identified as a key process related to rapid or even abrupt climate change (i.e., changes over a few years to a few decades). Improved understanding of this major component of ocean circulation and the ability to monitor, detect, and analyze changes in it could improve society's response to significant changes in climate.

Common among the societal themes is the need to develop the tools necessary to pursue research and to effectively translate the results of that research in ways that are useful to resource managers, policy-makers, and the general public. Society's ability to knowledgeably address key ocean-related societal issues depends on technological and intellectual innovation, incorporating many infrastructure assets, including ocean-observing and modeling capabilities. A healthy relationship between society and the ocean depends on having the scientific foundation to develop and implement new strategies to educate and instill a sense of stewardship in the public and translate research results into effective decision-making tools.

This document's final section details an implementation strategy that will ensure the ocean research priorities are successfully addressed. The implementation strategy describes characteristics of implementation, roles of various sectors, mechanisms for collaboration, the need for an infrastructure assessment, mechanisms for research translation, strategies for assessment and evaluation, and mechanisms for budget and plan updates. It does not stipulate specific federal-agency actions or budgets, provide timelines for activities, or mandate how other sectors should respond to the research priorities independent of federal collaboration—activities that are best defined in specific project plans. Carrying out this implementation strategy will require a national and international effort involving many sectors of the ocean community, from state, local, and tribal governance to academic institutions to nongovernmental organizations.

Vision for National Ocean Research Priorities

We envision a future in which our understanding of the world ocean, national coasts, coastal watersheds, and Great Lakes protects lives, enriches livelihoods, and enhances quality of life. At the same time, the research we undertake will help ensure the health and sustainability of our ocean ecosystem for years to come.

As expressed by the U.S. Commission on Ocean Policy (USCOP) and similar community efforts, increased scientific knowledge of the ocean will strengthen decision-making at all levels, enhance economic development, support sustainable resource use, and conserve biological diversity and natural beauty.

As the linchpin in this vision of the future, ocean research, bolstered by cutting-edge technology, will give us the ability to understand the past, observe the present, and predict the future. As an educated and informed nation, our knowledge will give us the tools to know well what the consequences and outcomes of our actions will be in the complex ocean ecosystems of which we are a part.

Introduction

The basic necessities for protecting lives, enhancing livelihoods, and improving quality of life depends, in no small measure, on functioning and healthy ocean ecosystems, the availability of ocean resources, and on a safe and secure maritime domain and associated marine-based economic sectors. The ocean[1] provides food, recreation, and other forms of enrichment, but at the same time, presents risks that, if not understood and respected, pose serious threats to lives and livelihoods. Informed management of the ocean environment will prevent or minimize the risks of illness and disease; bolster the resilience of coastal communities and regions to hazards that arise from the marine environment; enhance the national, homeland, and economic security of the nation; ensure the economic productivity of the open ocean, coasts, coastal watersheds, and Great Lakes; and improve the health of the ecosystems within them.

Understanding society's impact on the ocean and the ocean's impact on us forms the basis for ensuring a clean, healthy, and stable ocean environment that can be responsibly used and enjoyed by generations to come. This productive relationship depends on having the scientific foundation to develop and implement new strategies to enhance the health of the ocean, coasts, coastal watersheds, and Great Lakes; promote public health; and adjust resource management and use patterns. The ability to translate research results into effective decision-making tools, adapt governance structures to ensure conservation of resources, and educate and instill a sense of stewardship in the public will provide a path for a sustainable future. The purpose of this document is to develop and present ocean research priorities oriented around six societal themes that address key interactions between society and the ocean. If acted upon, these priorities will result in considerable strides toward enhancing the quality of life of the nation's citizens and safeguarding the health of the open ocean, coasts, coastal watersheds, and Great Lakes.

Community efforts, such as the USCOP, which inspired the activities to generate this plan, and the Pew Oceans Commission, produced detailed assessments of the state of the ocean and the means to address challenges revealed by these assessments. These assessments were developed using expert testimony, nationwide community input, and recommendations from previous reports. The ocean research priorities identified in this document are built on this foundation of community efforts to explore society's interaction with the ocean, to understand the impacts of human activities on the ocean, and to find new ways to restore and enhance the health of the ocean and its ecosystems.

This document considers the ocean as a dynamic global system; it does not attempt to identify research needs based on historical academic disciplines or independent activities such as resource distributions, fisheries assessments, and ocean currents. Research efforts must recognize the ocean's inextricable links to the land and atmosphere and take an interdisciplinary approach to exploring and understanding the ocean's role in many of society's most pressing challenges, such as understanding climate change and mitigating its impacts, sustaining natural resources, improving public and ecosystem health, and enhancing hazard mitigation. This document is also unique in that it recognizes the important relationship between society and the ocean and, therefore, places high emphasis on understanding the interactions between humans and ocean ecosystems—the human dimension of ocean issues. Finally, this document recognizes the unprecedented capabilities in the areas of advanced *in situ* and remotely sensed data acquisition, integration, and interpretation; information management; and computer simulation and visualization.

The national ocean research priorities described in this document build upon ocean science's strong legacy and take advantage of new interdisciplinary research approaches, sophisticated research and computational tools, and the availability of shared assets such as personnel and research platforms. An appropriately balanced research portfolio will provide insight into ocean processes that will enable better and timelier policy and resource-management decisions. These decisions will help ensure appropriate use and protection of the ocean, particularly the coastal margins, where so much of America lives, works, and plays, and that contributes so heavily to the nation's economy. This approach will require careful coordination among local, tribal, state, regional, and federal government agencies as well as academic, private sector, and nongovernmental entities. The ocean research priorities presented in this document are national in scope yet reflect the need to provide benefits at the local, tribal, state, regional, national, and global levels, while involving all parties in the enterprise.

Critical Elements

Three critical elements of science and technology will provide the United States with the scientific and technical means to redefine its relationship with the ocean for the better: developing the understanding and capability to forecast ocean processes; providing the scientific information needed to support ecosystem-based management, particularly in coastal and nearshore environments; and accelerating deployment of an ocean-observing system that will, in turn, advance both forecasting and adaptive ecosystem-based management capabilities. Aspects of these three key elements are evident throughout the national ocean research priorities identified in this document. These three elements are critical and necessary to expand society's vision of the ocean and ensure the ocean's legacy for future generations.

Developing the *understanding and capability to forecast certain ocean and ocean-influenced processes and phenomena* will change how society takes action in the future, much like weather forecasts do today. The tremendous potential of this capability has been demonstrated in the forecasts of El Niño and La Niña phenomena that are now used routinely by governments and the private sector. For example, these forecasts currently influence the buying and selling of energy resources and reduce loss by enabling agricultural interests to anticipate and prepare for El Niño-related weather hazards. These forecasts have emerged from investments in the basic understanding of equatorial Pacific Ocean circulation and its relation to weather, in observing systems capable of providing key data, and in model and simulation capability. Developing the capability to forecast key processes related to hurricane and severe-storm formation, ocean currents, fish stocks, and human-health hazards, to name a few, would expand not only the economic benefits of ocean forecasts, but their societal and environmental benefits as well.

Managing resources in ways that identify and account for the complex interactions between those resources and other components of the marine environment—including humans—is a substantial challenge. The principle of ecosystem-based management[2] is now widely recognized as one of the most effective ways to cope with a variety of increasing natural and human-induced pressures. Providing *the scientific support for ecosystem- based management* requires a multi-dimensional, multidisciplinary effort to enhance current understanding of ecosystem processes, determine which interactions are most critical, and assess the dynamics of the natural and human factors affecting those interactions, including how those factors will change in the future.

Deployment of a robust ocean-observing system that can describe the actual state of the ocean, coupled with a process to synthesize observational data, will fundamentally alter society's view of the ocean environment. Observations underpin fundamental knowledge of the open ocean, coasts, coastal watersheds, and Great Lakes. Although much work remains, communities interested in ocean research and management have developed mature plans for many components of an integrated, global ocean-observing system. Deploying the priority elements of that observing system will increase society's access to the ocean and allow researchers to enable the promise of ocean forecasting and ecosystem-based management during the next decade

EXPANDING THE SCIENTIFIC FRONTIER: THE NEED FOR FUNDAMENTAL SCIENCE

Scientific discovery, driven by competitive, peer-reviewed investigations, is the foundation of the nation's research enterprise and is an intrinsic and highly valued component of the ocean research portfolio. Fundamental research that expands the scientific frontier enhances and deepens our understanding of the ocean and its role in the Earth system. It is essential that the nation cultivate and investigate new ideas about the ocean and new approaches for exploring the marine environment that may challenge existing interpretations. In doing so, society should recognize and even encourage risk-taking in supporting the most exciting and promising ideas for making progress in understanding the ocean. Progress requires the continued support of both systematic measurements of the ocean's properties and the freedom to pursue new ideas and technology. Discipline-oriented research focuses on examining the intricacies of one facet of science, while interdisciplinary collaborations capitalize on this focus to investigate connections and feedbacks between diverse elements. Both of these efforts need to be enhanced in parallel to enable comprehensive understanding of the ocean and its place in the Earth system. The path ahead as presented in this document necessarily includes the need for "creative individuals to pursue the kind of fundamental scientific research that can lead to unforeseen breakthroughs"[3].

The ability to access all aspects of the ocean environment, made possible by fostering scientific and technological innovation, will enable breakthroughs in basic understanding of ocean biology, chemistry, geology, and physics and the connections among these disciplines. Improved access to the open ocean, coasts, coastal watersheds, and Great Lakes depends on advances in infrastructure and technology, from advanced sensors to satellites and unmanned vehicles. The development of innovative tools, from remotely operated and autonomous vehicles, to molecular techniques and genetic sequencing, to physical, chemical, and biological sensors, will facilitate novel experiments and permit the study of processes ranging from isolated episodes to global cycles. Observing and understanding ocean processes that operate on different temporal and spatial scales and in different regions requires a balanced combination of *in situ* observing platforms, a robust fleet of ships, remotely operated autonomous vehicles, satellites, and land-based marine laboratories supporting specialized equipment and instrumentation. The reach of ocean science research can be significantly expanded by leveraging the capabilities of other disciplines, including nanotechnology, genomics, and robotics, enabling new access to and perspectives on the ocean environment. A workforce eager to push the limits of scientific knowledge and technological innovation will support the scientific discoveries necessary to address fundamental scientific questions and to advance the use of scientific research for operational activities to help meet national needs.

The research community identifies many promising areas for scientific investment through workshops and other planning activities. The National Research Council and other community groups also publish analyses of promising new directions in fundamental ocean research that are used by funding agencies to guide investments[4]. Attempting to prioritize research efforts driven by new ideas and the desire for discovery would constrain these creative and visionary activities critical to the research enterprise. By definition, unforeseen breakthroughs and paradigm shifts cannot be planned. Therefore, this document focuses on underscoring—rather than attempting to define or enumerate comprehensively—the fundamental research efforts that provide the foundation for understanding the ocean. This document emphasizes the research efforts with particular anticipated societal applications, while invoking the fundamental research that provides the foundation for those applications.

FOCUSING THE NATION'S OCEAN RESEARCH ENTERPRISE

Framing the Approach

The scope of ocean science is to promote *discovery* that will provide new insights and perspectives on the ocean environment; impart greater *understanding* of ocean properties and processes; and expand opportunities to *apply* that knowledge to enable safe, prosperous, and sustainable use of the ocean. As the three interlinked components of the research enterprise, discovery, understanding, and application serve as the foundation for the research priorities developed in this document.

Discovery

Discovery entails both the pursuit of the unconstrained fundamental science questions (see Expanding the Scientific Frontier) that drive the imagination and the exploration of new phenomena and terrain, remote geographies, and unique marine systems. Discovery provides the foundation upon which society builds comprehension of life; it has the potential to fundamentally shift our understanding of the ocean, its role in the Earth system, and society's place in that system.

Understanding

An understanding of ocean processes is required to predict the consequences of interactions between humans and the ocean, and manage those interactions. Supported by discovery efforts, the need to understand often focuses on those processes that cascade through spatial and temporal scales, describe the interactions between ecosystem components, and/or cross thematic boundaries. Enhancing understanding is an ongoing task that requires continuous innovation in both ideas and technology. The benefit of this understanding is that new knowledge can then be applied to promote responsible use and stewardship of the marine environment.

Application

A critical component in the research process is to apply the knowledge gained from discovery and understanding to serve the public interest. Application includes efforts to translate information so that policy-makers, managers, and the public will be able to understand and use the knowledge to make informed decisions as well as provide feedback to define additional research needs. Application also includes developing new tools, new techniques, and new approaches to address complex issues and inform societal choices.

Identifying Ocean Research Priorities

This document focuses national attention on six themes that represent key areas of human interaction with the ocean.

- Theme 1: Stewardship of Natural and Cultural Ocean Resources
- Theme 2: Increasing Resilience to Natural Hazards
- Theme 3: Enabling Marine Operations
- Theme 4: The Ocean's Role in Climate
- Theme 5: Improving Ecosystem Health
- Theme 6: Enhancing Human Health

These themes and associated research priorities were developed through collaborative efforts involving all federal agencies with interest and responsibilities linked to the ocean, as well as representatives from state and local governments, academia, industry, and non-governmental organizations. For each societal theme, priorities are identified that *focus on essential research efforts that must be undertaken in the coming decade* (Figure 1). The research priorities encompass natural sciences as well as those disciplines focusing on human interaction with the ocean, such as economics and other social sciences (e.g., sociology, anthropology) and public health. In addition, tools universally critical to pursuing the research priorities and realizing their results—ocean-observing and modeling capabilities, mechanisms to translate research results into effective decision-making tools, and initiatives to educate and instill a sense of stewardship in the public—are described and revisited throughout this document.

The ocean research priorities emerged from a larger suite of questions that were considered, and emphasize topics such as systems research and predictability. The following questions, with equal overall weight, were used to identify the most compelling research priorities for each theme, recognizing that the prioritization criteria for one theme may not be equally applicable to another:

- Is the proposed research transformational (e.g., will the proposed research enable significant advances for insight and application, even with potentially high risk for its success; would success provide dramatic benefits for the nation?)?
- Does the proposed research impact many societal theme areas?
- Does the research address high-priority needs of resource managers?
- Would the research provide understanding of high value to the broader scientific community?
- Will the research promote partnerships to expand the nation's capabilities (e.g., contributions from other partners, including communities outside of ocean science, such as health science; unique timing of activities)?
- Does the research serve to contribute to or enhance the leadership of the United States in ocean science?
- Does the research contribute to a greater understanding of ocean issues at a global scale?

- Does the research address mandates of governing entities (federal agencies, state, tribal, and local governments)?

Four near-term priorities were developed to initiate rapid progress towards the 20 ocean research priorities. These near-term priorities are not a direct subset of the larger research priorities, but incorporate issues highlighted in many of them. These priorities were selected from a larger suite of efforts using the criteria above, with an added focus on impact (i.e., the value of the work), urgency (i.e., the need for a concentrated effort over the next two to five years), and partnerships (i.e., the effort will maximize collaborations among agencies and external partnerships).

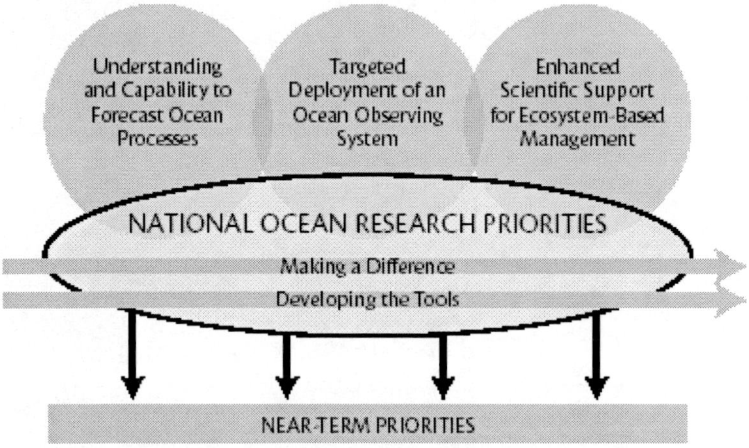

Figure 1. Overview of Ocean Research Priorities.

Societal Themes

Fishery survey catches document the abundance and recruitment trends of species. This catch of Acadian redfish was taken in the gulf of maine aboard the nOAA R/V *Albatross IV*. *Photo credit: NOAA.*

The ocean is home to some of the nation's cultural resources. *Photo credit: NOAA*

Energy exploration and production from the ocean represent important contributions to the nation's energy supply. *Photo credit: Department of the Interior.*

THEME 1: STEWARDSHIP OF NATURAL AND CULTURAL OCEAN RESOURCES

The resources of the open ocean, coasts, coastal watersheds, and Great Lakes generate tremendous benefits and opportunities. The ocean is a source of food, minerals, and energy and is used for transportation, recreation, and tourism. The ocean preserves a record of the nation's cultural past. The ocean remains a vast, unexplored realm with the capacity to provide new pharmaceuticals, industrial products, and energy sources. At the same time, its resources are subject to many pressures, such as overfishing, habitat destruction, and competition with invasive species. To unlock the full resource potential of the open ocean, coasts, coastal watersheds, and Great Lakes, society must improve understanding of these resources and balance their health and use.

Rationale

U.S. waters are the source of many of the resources society uses daily. The United States has the largest Exclusive Economic Zone (EEZ) of any nation[5]. Fishery production within the EEZ and contiguous inshore waters supports a $60 billion annual seafood industry and a $20 billion recreational fishing industry[6]. Recent advances in aquaculture are providing an increasing contribution to the seafood supply. The total current market value of the oil and gas inventory is approximately $8 trillion, consisting of $1 trillion in discovered remaining reserves and an estimated $7 trillion in undiscovered, technically recoverable resources[7]. Estimates for offshore wind energy suggest a potential of 900 gigawatts within 50 nautical miles of the coast[8]. Approximately 200,000 trillion cubic feet of gas hydrate are present on land and the U.S. outer continental shelf (OCS)[9,10]. The Great Lakes are the largest source of freshwater in the world[11]. Healthy ocean and coastal natural and cultural resources provide the foundation for a huge coastal tourism and recreation industry that is the fastest growing area of the ocean economy[12]. Prehistoric landscapes, shipwrecks, and historic and living waterfronts along the nation's coasts and Great Lakes all contribute to the national cultural heritage.

The scale and diversity of ocean resources is immense, however, resource use and development often compete with other societal needs and values. Overexploitation of many fish stocks and alteration of habitats due to human activities and/or natural influences such as climate change have had profound consequences for many ecosystems and fishing communities. Increasing domestic energy production from the EEZ (renewable and non-renewable) brings with it concerns about environmental impacts. Increasing populations at the coasts and Great Lakes impact freshwater supplies and inappropriate wastewater disposal can degrade water quality, impacting marine resources. Evaluating and addressing environmental impacts of resource use and extraction, combined with increased understanding of the factors influencing overall ecosystem health, can help balance the pressures being placed on coastal ecosystems, enable restoration of degraded habitats, and, ultimately, support robust and coordinated ecosystem-based management and governance strategies for sustainable resource use.

Research Priorities

Central to effective management of natural and cultural resources is the ability to accurately assess their current condition. This knowledge will provide the foundation to understand the complex relationships between living and non-living resources and the cumulative impacts of human activities, and help determine the likely impacts of various management alternatives. Research into issues of resource development, use, and extraction, such as exploration for mineral resources and pelagic and benthic communities, and regional variability and influence on resources, will help society prevent major impacts to ecosystems, promote sound development and use of resources, preserve cultural sites, and support management efforts to restore depleted populations to healthy and sustainable levels.

Research Priority 1: Understand the status and trends of resource abundance and distribution through more accurate, timely, and synoptic assessments.

Assessing the impacts of resource use and extraction (e.g., fisheries, ocean mining, tourism [cultural sites]) requires measuring the abundance and distribution of living and non-living resources in the open ocean, coasts, coastal watersheds, and Great Lakes. Capabilities necessary for these measurements include the ability to: assess fish-stock and protected-resource status and health; monitor living resources (spanning multiple trophic levels) nearly continuously over wide swaths of ocean at appropriate levels of species resolution; assess the spatial and temporal variability (both natural and use-induced) of resources (e.g., biota, energy, minerals, and pharmaceuticals), including in deep-water settings; and provide long-term and sustained monitoring and mapping of natural and cultural resources. Development and implementation of these capabilities will enable adaptive approaches to the management of natural and cultural resources, using empirical evidence to design and modify management efforts (e.g., individual harvest quotas, special use areas, marine protected areas [MPAs]) to be most effective.

Research Priority 2: Understand interspecies and habitat/species relationships to support forecasting resource stability and sustainability.

Understanding the individual and cumulative impacts of an ever-widening array of human activities (e.g., transportation, recreation, land use, fishing, energy exploration and development, and increased food production from aquaculture) on water and habitat quality and the ocean's ability to support marine life and biodiversity is necessary for enabling ecosystem-based management approaches to ocean-resource use. Often the data necessary for these assessments are difficult to obtain because of practical limitations of controlled experimentation and replication in natural systems. Investing in observations, process studies, and advanced modeling will expand current understanding of impacts at appropriate temporal and spatial scales, and help identify crucial data and process-understanding gaps such that the proper resource-management techniques can be developed and implemented. Priorities to enable improved natural-resource forecasting include: (1) development and validation of ecosystem and species interaction models at appropriate scales that incorporate feedback mechanisms among trophic levels; (2) assessments of regional and local models of environmental changes (e.g., responses to annual, decadal, and longer-term climate drivers; rapid regime shifts; hydrodynamic circulation; watershed discharge) to help determine how these changes impact resources; (3) development of approaches and scenarios to understand and integrate the specific and cumulative impacts of various natural resource policies on living resources and human communities (incorporating improved assessments of impacts on ecosystems [see Improving Ecosystem Health]); and (4) collection of necessary data (observational and experimental) to support model robustness.

Research Priority 3: Understand human-use patterns that may influence resource stability and sustainability.

Determining the "worth" of natural and cultural resources and evaluating effects of alternative management scenarios requires considering economic (i.e., market and non-market valuation), sociological, and cultural factors, and potential competing uses, in addition to ecological factors. These assessments will also support evaluations of the socioeconomic trade-offs inherent in management efforts. Human-use data can be acquired through expanded surveys and analysis of economic and sociological factors associated with a broad range of open ocean, coast, coastal watershed, and Great Lakes resource activities. The use of geocoded socioeconomic data (e.g., resource cost and earning, demographics, income), combined with new methods to assess non-market valuation (see Improving Ecosystem Health), will enable the integration of socioeconomic and natural processes into models to determine interactions between these systems. Human behavior and economic research, combined with resource-use assessment and modeling, will help support effective evaluation of the success of various management options on resource use and sustainability.

Research Priority 4: Apply advanced understanding and technologies to enhance the benefits of various natural resources from the open ocean, coasts, and great lakes

Key technological developments are required to facilitate long-term and effective resource management; to make more responsible use of available sustainable and non-sustainable resources, such as energy, minerals (including sand), and pharmaceuticals; and to support alternatives to resource extraction (e.g., aquaculture, alternative energy sources), in conjunction with efforts to enhance the efficacy of and limit impacts from marine operations. Areas of focus include, but are not limited to: development of sustainable approaches to aquaculture that consider implications for surrounding ecosystems, wild genetic resources, and impacts on coastal economies; advancement of new sustainable energy technologies from the ocean, including efficient methods for power generation (e.g., wind and tidal power), and scientific information to fully consider impacts on marine life; develop-ment of new generations of military subsurface detection systems or other mitigation methods that are less deleterious to species sensitive to acoustic emissions in certain sound ranges; development of bycatch-reduction technologies for fisheries and protected resources (e.g., seabirds), enabling sustainable resource use; and use of advances in exploration technologies (e.g., geospatial, biooptical) to map and characterize the EEZ and the U.S. continental shelf, to establish the outer limits of the U.S. continental shelf where it extends beyond the limits of the EEZ, and to assess its important ecological and economic resources.

Necessary Tools

Technologies and procedures will be needed to enable rapid, efficient, and synoptic assessment of ocean resources (including inventories and alterations). An expanded and technologically more capable domestic fleet of ships would improve the capability to provide important information on the status of managed populations and ecosystem effects of human populations. Additionally, improved remote-sensing tools are needed to obtain necessary biological (e.g., biomass) and physical (e.g., current direction, wave height) parameters, and to survey deep waters, particularly as energy exploration moves to the edges of the OCS. Advances in unmanned aerial and autonomous vehicles will aid in collecting synoptic resource information. In addition to monitoring and assessment with existing technologies, new and improved technologies and protocols are needed to expand the capabilities of fixed and mobile observing platforms to include biological and chemical sensors, and advanced acoustics and mapping capabilities.

Enhanced information-technology and data-support infrastructure are essential. Comprehensive environmental databases with appropriate spatial and temporal resolution that integrate information from local, regional, national, and global sources are key to understanding ecosystem interactions and providing the necessary information for effective and adaptive resource management. Information systems using geographic information systems (GIS) can integrate diverse data sets (e.g., physical, biological, social, cultural) to promote spatial analysis and static and dynamic modeling. Development of integrated ecosystem assessments will, in turn, expand understanding of interrelationships among the physical environment, ecosystems, and human activities.

Implementation of these new approaches requires more technical and scientific personnel skilled in diverse disciplines, including natural sciences, social sciences, and information technology for the operation, maintenance, and interpretation of information. Investments in training and maintaining this workforce are key priorities for next- generation, natural-resource decision-support systems, improved resource management, and decision-making.

THEME 2: INCREASING RESILIENCE TO NATURAL HAZARDS

Recent hurricanes and tsunamis have clearly demonstrated the potential for natural disasters to have economic, environmental, social, and public-health impacts on regional, national, and global scales. Communities, maritime operations, cultural resources, social services, and ecosystems are vulnerable to coastal and marine physical hazards. Although society cannot eliminate natural hazards, their impacts can be reduced. Sustained and coordinated investment in research and technology will provide the knowledge and information base needed to assess and reduce risk, save lives and property, ensure more rapid recovery and effective mitigation, and develop informed and effective responses to future hazard events.

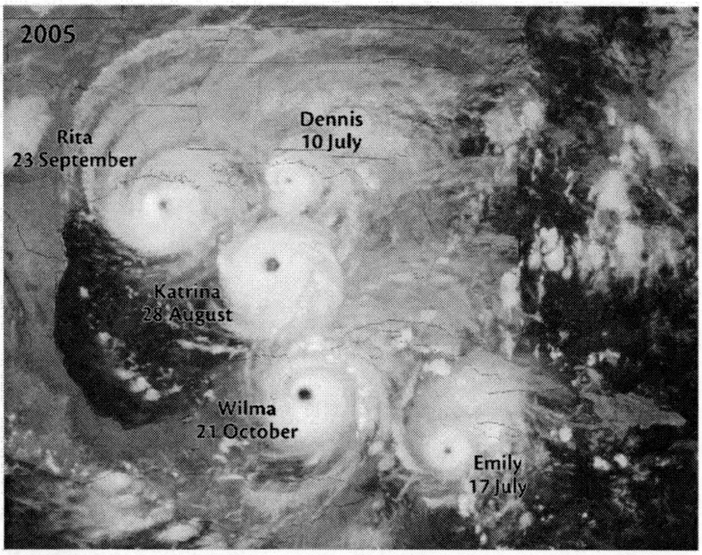

Understanding the impact of severe storms on coastal regions is critical to improving hazard resilience. *Composite photo: University of Wisconsin CIMSS Tropical Cyclones Group, NOAA, NASA.*

Rationale

With over 50 percent of the U.S. population living in coastal counties and that number rising each year[13], the United States must be adequately prepared for coastal disasters. Even as the nation continues to direct significant federal funding toward science and technology to assist in reducing the impacts of natural hazards, the United States still faces enormous losses each year from ocean and coastal hazards, including:

- Severe storms, hurricanes, and tornados, and the associated coastal and offshore wind, wave, and current damage
- Coastal inundation and flooding from storms, tsunamis, and regional meteorological events
- Earthquakes, landslides, and volcanic eruptions along the coast and offshore, and resulting tsunamis

The costs of these events are significant and will likely increase. In 2005, Hurricane Katrina devastated the Gulf Coast and is estimated to be the costliest natural disaster yet to strike the United States[14]. The potential short-term and long-term human and economic costs, including impacts to public health, threats to subsistence, disruption of livelihoods, population relocation, infrastructure damage, and ecosystem degradation, to at-risk regions from tsunamis (Pacific Northwest, Alaska, Hawaii, southeastern United States, Caribbean) are substantial.

Effective response to natural hazards is impeded by limitations in the ability to comprehensively assess hazard potential, vulnerability, and risk. Human and natural processes (e.g., land development, sea-level rise) constantly alter system vulnerability and the risk of hazards to coastal communities. To reduce the social and economic impact of natu-ral hazards, the public and decision-makers must be educated about the risks, associated impacts, and response costs of natural hazards. A sound scientific and technological basis for decision-making must include improved models of hazard impacts, more accurate and timely forecasts, and comprehensive assessments of future hazard potential, vulnerability, and risk.

Research Priorities

Hazard assessments must be comprehensive, reflecting multiple hazards and their impacts on diverse economic, social, and ecosystem sectors. Physical processes must first be assessed and then coupled with improved understanding of the role of social systems and human behavior in increasing vulnerability or enhancing resilience. This effort will be enhanced by investigations of ecosystem health (e.g., in response to resource and land use), impacts of climate change (e.g., sea-level rise, storm intensification and/or concentration), and direct and indirect risks to public health that broadly influence hazard resilience. Assessment of hazard impacts and risk that incorporate improved hazard understanding, effective research translation, communication, and education, are integral for developing a "risk-wise" population, supporting cost-effective strategies designed to increase resistance, enhancing resilience to hazards, and promoting avoidance.

Research Priority 5: Understand how hazard events initiate and evolve, and apply that understanding to improve forecasts of hazard events.

Quantifying future hazard potential requires understanding hazard generation (e.g., storms, submarine and coastal landslides, tsunamis, flooding) and past hazard occurrence. Enhanced hazard forecasts, particularly of storm formation, track, intensity, and associated waves, surge, and flooding, will support more effective responses to developing hazard events. Research in this area should focus on process studies and development and validation of models of hazard generation (e.g., tsunami-source modeling, seafloor-stability modeling, storm formation) and evolution (e.g., tsunami propagation, storm and inundation modeling). Probabilistic models and assessments of hazard potential should include effects of land subsidence and future climate change (e.g., changes in storm intensity or frequency, sea-level rise, landscape change) on hazard potential and vulnerability.

Research Priority 6: Understand the response of coastal and marine systems to natural hazards and apply that understanding to assessments of future vulnerability to natural hazards.

Natural hazards impact infrastructure directly through alterations to the underlying landscape and through secondary processes (e.g., slope failures, shoreline change, inlet formation, coastal erosion, sediment transport, flooding). Natural hazards have significant impacts on coastal features, such as shorelines, as well as cascading and nonlinear impacts throughout ecosystems, during and after the hazard event. Natural systems, such as wetlands and reefs, play a significant role in coastal resilience. Understanding the capacity of these systems to mitigate the effects of natural hazards, how altering the system (i.e., through physical destruction, sediment diversion) affects this capacity, and understanding altered ecosystem dynamics and processes, will improve the ability to facilitate recovery of ecosystem functions and to undertake appropriate restoration efforts. Priorities include understanding and modeling landscape change (including impacts/influences across the adjacent watershed) associated with coastal hazards, determining structure and infrastructure resilience, and assessing vulnerability of coastal communities, public health, infrastructure, marine operations, and ecosystems. Models based on observations of pre-event conditions and post-event impacts are required to forecast the magnitude and nature of impacts, the changing vulnerability of the altered landscape, such as the influence of land use and restoration on a system's ability to resist or adapt to hazards, and the long-term impacts on ecosystem functions and health.

Research Priority 7: Apply understanding to develop multi-hazard risk assessments and support development of models, policies, and strategies for hazard mitigation.

Integrating observations, data, and models will enhance understanding of hazard potential and vulnerability, and improve risk assessments for coastal communities, public health and safety, infrastructure, marine operations, and ecosystems. Research efforts must identify vulnerable ecosystem functions and infrastructure components, determine the potential for cascading component failure, and assess the efficacy of natural (e.g., bar-rier islands, coastal wetlands) and engineered systems (e.g., hurricane barriers, levees) in hazard mitigation. These assessments should be integrated with improved understanding of system responses to hazards and incorporated into comprehensive risk assessments. These risk assessments will support identification of effective and affordable systems, materials, and technologies for hazard-resilient and resistant communities, infrastructure, and ecosystems as well as support development and management approaches minimizing hazard impacts.

Models of resilience and resistance to hazards must also address social and economic drivers that influence pre-hazard activities (e.g., land use, shoreline development) and responses to hazards (e.g., warning responses, community development, economic restructuring). Integration of models and risk assessments into decision-support tools will inform short- and long-term public and policy responses to hazards that will minimize economic, social, and environmental costs associated with natural hazards. Integrating natural and social sciences will ensure that hazard products and policies address public perceptions of risk, communicate the consequences of community and individual responses to immediate and long-term hazards, and enhance appropriate public response to hazard assessments and warnings at all levels ("risk-wise" behavior).

Necessary Tools

Meeting the broad requirements necessary to reduce hazard impacts and increase hazard resilience will require continued expansion of existing observational systems, data-delivery systems, and modeling capabilities at regional, national, and global scales. Sustained and integrated marine and coastal observations, including technological improvements in remote and *in situ* sensing of multiple time-sensitive parameters (oceanographic, geophysical, hydrological, chemical, biological, geographic) are required. Comprehensive geospatial characterization combined with sustained and reactive deployment modes of operation are required to acquire baseline (pre-impact) and post-hazard data. Deployments must be capable of assessing diverse hazard processes and impacts, such as coastal and inland inundation and water quality. These deployments would support development of improved models of, for example, storm evolution and coastal response. The capability to provide up-to-date characterization of coastal vulnerability and pre-storm conditions (including topography and bathymetry, land-use/land-cover, monitoring of riverine and coastal water levels) must be maintained.

Multi-hazard assessments will require information infrastructure supporting development of an "all-hazard" GIS resource with defined standards supporting information integration and application. Advanced computational resources and the establishment of community computational standards are required for data-assimilation systems, coupled land-ocean-atmosphere models, and data management and delivery systems. The design and implementation of robust, reliable, and widely available notification systems will help enable more effective communication of warnings.

Research communities that address hazards currently focus on single hazards, single elements of the hazard cycle (e.g., generation, impact, policy), or single research disciplines (e.g., physical, ecological, social, and engineering sciences). These diverse communities must be coordinated and integrated to inform and evaluate mitigation strategies with the broadest consideration of costs and effectiveness (i.e., on economic, sociological, ecological levels). An integrated approach will lead to more effective translation of research results and development of tools for informed and comprehensive decision-making.

Estin, FL, August 24, 2005. Coast Guard boat crews prepare for search and rescue missions. *Photo credit: Petty Officer 2nd Class NyxoLyno Cangemi, USCG*

Maritime transportation is a critical element in U.S. commerce. *Photo credit: Bill Snyder*

THEME 3: ENABLING MARINE OPERATIONS

Marine operations are essential components of the global economy and vital elements of economic prosperity, national security, and homeland defense[15]. Marine operations require freedom of navigation in the global ocean and the ability to conduct mapping and charting activities, gather tide and current information, and use U.S. ports, harbors, estuaries, and the Great Lakes. Marine operations also deal with issues such as bridge clearance, dredging, navigation aids, and ice coverage. Forward-thinking, innovative research, coupled with technological advances, will permit marine operations to meet challeng-ing requirements for increased levels of transportation and commerce in the maritime domain and to address security concerns, while balancing sustainable use and protection of the environment.

Rationale

Marine operations comprise a broad set of activities, including merchant ship traffic, commercial fishing, marine public transportation, and military operations that occur in the open ocean, coasts, coastal waterways, and Great Lakes. Even a brief interruption in marine operations can have serious consequences for many sectors of society, national security, and the U.S. economy. The broad scope and far-reaching impact of marine operations requires expanding knowledge in several key areas. Though the priorities identified in this theme emphasize the transportation sector, the knowledge derived from these efforts will have broad applicability in other operational areas, such as national security (see Box 1).

Given that the majority of the nation's commerce (by weight) travels through U.S. ports and is anticipated to expand, including increased volume of open-sea shipping (potentially through an ice-reduced or ice-free Arctic[16]) and short-sea shipping[17], the capacity of the U.S. marine transportation system must be enhanced[18]. With that expansion, projections state that trade will increase, ships will become larger, and population along coasts will grow, all of which will have impacts on the marine environment and the nation's ability to conduct marine operations.

The United States has more than 90,000 miles of shoreline, 25,000 miles of navigable waterways, and 3.4 million square miles of open water within the U.S. EEZ[19,20]. Safe and effective navigation of harbors, ports, and waterways by various users requires reliable weather and sea-state forecasts and up-to-date maps and charts.

Numerous other commercial but non-transportation uses of the maritime domain exist, including offshore aquaculture, energy extraction/harnessing (i.e., facilities for oil, gas, and wind), seabed and mineral-rights uses, and laying of subsea communications and cables, all of which greatly affect the environment and use of the maritime domain. Resolving conflicts among the numerous existing and emerging uses of the maritime domain represents a challenge. To properly address these issues and to use the results of the research outlined below effectively requires improved communication and collaboration among the diverse stakeholders involved in marine operations, including industry; local, tribal, state, and federal governments; and researchers.

Research Priorities

Increased understanding of the environmental impacts of marine operations and the environmental conditions affecting marine operations is necessary as these activities expand (e.g., increased oil and gas development, aquaculture, transportation, and other ship operations in the open ocean and close to shore [e.g., military, search and rescue]). This increased understanding is necessary to ensure a balance of operations and environmental protection. Additionally, more effective collection, processing, and modeling of environmental data are required to ensure safe and secure operations.

Research Priority 8: Understand the interactions between marine operations and the environment.

Central to enabling effective and environmentally viable marine operations is understanding how marine operations (including port development and vessel maintenance and operations) affect the environment, and how environmental conditions affect operations. This understanding will be particularly important in environmentally sensitive areas such as the Great Lakes, coastal areas with low tidal exchange, and coral-reef systems. Areas of study include the release, dispersion, cycling, and cumulative ecological impacts of contaminants (i.e., from oil spills and releases, air emissions); interactions with marine life (e.g., bird migration, ship strikes, ocean sound); impacts of aquaculture development and production; factors contributing to the introduction and persistence of invasive species (e.g., ballast water); and effects of expanding transportation routes (e.g., Arctic routes). These research efforts will build upon investigations into factors influencing ecosystem dynamics (e.g., invasive species, contaminants) and sustainable resource use (e.g., impacts on wild stocks from aquaculture). They will also enable the development of mitigation strategies and technologies (e.g., passive sonar, spill prevention and containment, ballast- water treatment, vessel routing to avoid ship strikes).

Research is also needed to address the feedback effect of environmental conditions on marine operations. Such research includes refining sediment transport models to enable rapid, efficient, and environmentally sustainable dredging and dredged-material management. Investigations of environmental changes related to climate change (e.g., sea-level rise, sea-ice abatement[21], lake-level decreases) will help determine impacts on the development of alternative transportation routes, necessary changes to ports and harbors, and the stability and safety of energy-extraction platforms.

Research Priority 9: Apply understanding of environmental factors affecting marine operations to characterize and predict conditions in the maritime domain.

Enhancing environmental observation, characterization, and forecasting of ocean and waterway conditions (e.g., currents, turbidity, surface waves, sea-ice extent, lake levels, biogeochemical conditions) across the global ocean (open ocean, coasts, coastal waterways, and Great Lakes) is necessary for safe and efficient marine operations. Increased precision in forecasting marine conditions (e.g., improved real-time, ocean-current models; storm- surge projections, in conjunction with efforts to address natural hazards) will promote safe and efficient operations, including transportation, military operations, and search-andrescue efforts, and will minimize negative impacts on marine operations and the environment. Technologies (e.g., robust sensors and infrastructure, autonomous vehicles) must be developed to enhance data collection in all weather conditions to support high-spatial-resolution and near-real-time forecasting throughout the open ocean and coastal zone.

Research Priority 10: Apply understanding of environmental impacts and marine operations to enhance the marine transportation system.

As commerce increases, the capacity of the marine transportation system must also increase through improved waterway traffic mobility and safety, enhanced port operations and productivity, expanded capabilities (e.g., short-sea shipping, higher-capacity vessels), and reduced congestion. Effective seaport planning, including port development, expansion, or modification and coordination with land and air transportation routes, and waterway improvements, must incorporate environmental impacts, social and economic drivers, and operational needs. Social and economic assessments that address human demographics, land use, income, and output must be developed and incorporated into risk assessments and models that address environmental impacts of marine operations (e.g., benthic disturbance, contaminant releases, impacts from invasive species) and operational requirements (e.g., port and waterway depth and capacity). These models can be used to design the safest and most effective and environmentally sustainable mechanisms to expand the operational capacity of the marine transportation system.

BOX 1. ENSURING NATIONAL AND HOMELAND SECURITY

The coast and open ocean are critical domains for the security of the United States, both at home and abroad. National-security operations in the ocean take place globally and often require continuous, near-real-time monitoring of environmental conditions using tools such as autonomous sensors, targeted observations, and adaptive modeling. Expanded awareness of the ocean environment for security operations requires improved capabilities in:

- Autonomous monitoring of desired ocean parameters in any location for extended time periods
- Integration of multi-sensor data, including data from a robust, satellite-based global earth observation system
- Numerical models to provide nowcasts and forecasts for critical parameters.

These capabilities, combined with improved understanding of the ocean environment enabled by other ocean science research activities, will support accurate ocean-state assessments and allow future forces to conduct joint and combined operations in nearshore and deep-ocean operating environments, anywhere and at anytime.

Necessary Tools

Achieving safe, effective, efficient, and secure marine operations that support environmental protection requires a diverse suite of infrastructure and technology, ranging from assessment methodologies (e.g., rapid-assessment methods for detecting marine contaminants/pollutants and harmful, non-indigenous species) to observing systems and information networks. Providing accurate and comprehensive environmental information will require expanding observational networks to monitor, record, and present real-time, surface-monitoring data (e.g., high-frequency, coastal-based radars). This expansion will require advancing sensor and technology development, particularly for autonomous and persistent observations, as well as for long-term observing systems; expanding real-time or near-real-time data collection on environmental variables by incorporating observational capabilities of ships of opportunity (e.g., fishing, cargo, and passenger vessels); and enhancing automated and autonomous bottom-mapping capabilities for change detection to improve rapid, full-scale survey scheduling.

Data collected by the observing systems must be accessible through a comprehensive national data network, either through a single system[22] or a distributed network. Developing this data network will require new methodologies that address gaps in data collection, sharing, and interoperability of technologies, and should permit integration of existing research into operational systems (e.g., systems providing real-time navigation data to vessels). This data network should be able to link with other databases, such as those focusing on ecosystem data, and developed in accordance with international standards for data exchange. The national data network will also provide the data needed for models simulating multiple scenarios to better understand potential impacts weather events or man-made disruptions on marine operations, and to support operations restoration plans.

THEME 4: THE OCEAN'S ROLE IN CLIMATE

The ocean plays a fundamental role in governing climate through its capacity to store and distribute heat and carbon. The challenge is to accurately assess the ocean's past and present state, processes, and phenomena influencing climate, and society's influence on them, and to improve predictions and projections of climate change. These predictions and projections will improve society's ability to respond to and reduce, where feasible, climate- related hazards; to adapt to climate change and variations (e.g., sea-level rise, changing weather patterns); and to inform management and policy decisions addressing human and environmental impacts.

Rationale

The ocean covers more than 70 percent of the planet and has a much higher capacity to store heat than the atmosphere. Imbalances of the planetary energy budget are manifested in changing ice conditions and ocean temperatures. Runoff from melting ice sheets and glaciers and an expanding water mass due to increasing temperatures lead to rising sea levels. Low-lying coastal regions are particularly impacted by gradual sea-level rise in addition to episodic events (e.g., storm surges, coastal flooding). A more complete understanding of how changing sea level impacts coastal communities is needed at regional, state, and local levels.

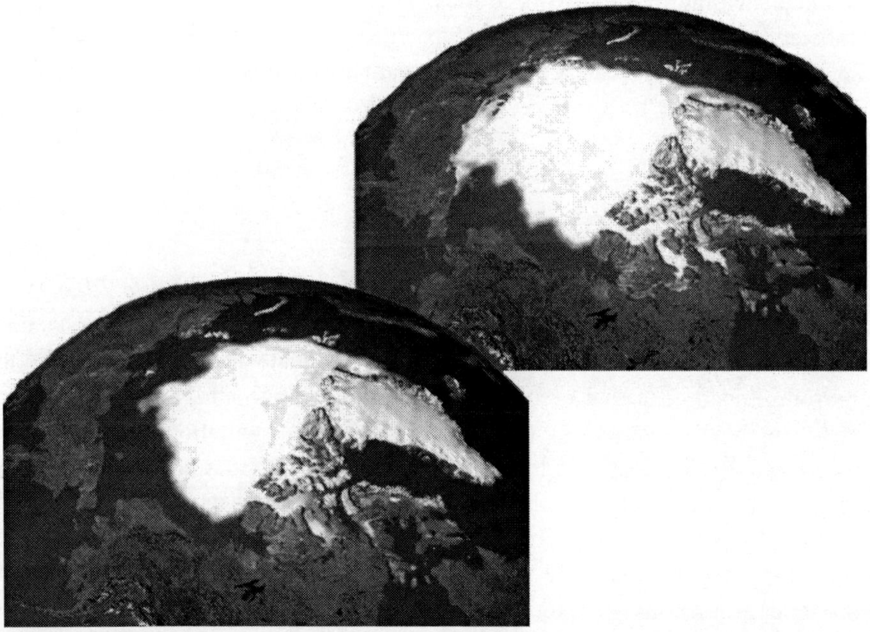

A comparison of september minimum sea ice concentration over the Arctic circle, acquired in 1979 (top) and 2005 (bottom). *Image courtesy of Scientific Visualization Studio, NASA Goddard Space Flight Center.*

The tropical ocean is a vital component of seasonal-to-interannual climate variability. For example, the El Niño-Southern Oscillation is known to have a substantial impact on many regions of the world, influencing agricultural yields and hurricane intensity. Current evidence suggests that the tropical ocean also influences the occurrence of multi-year droughts. The global ocean is also an important component of the global water cycle; it receives and redistributes freshwater from rivers and ice discharge, and provides moisture (through evaporation) to the atmosphere that precipitates over the ocean and continents. Alterations in large-scale ocean circulation can influence long-term climate change and, possibly, short-term or abrupt changes.

Recently, changes in major ocean processes have been demonstrated to affect marine ecosystems, causing, for example, large swings in the populations of commercial fisheries, changes in seabird-population distributions, and coral-reef-bleaching events. Warming oceans have resulted in changes in the distribution and abundance of populations (e.g., at the poles, in temperate regions, and in coral-reef habitats). The future amount of greenhouse gases in the atmosphere, such as carbon dioxide (CO_2) and methane, will depend, in part, on the exchange of these gases in open-ocean and coastal systems (e.g., wetlands). Substantial changes in ocean chemistry as a result of oceanic uptake of these gases will further influence ecosystems and their processes. For example, rising atmospheric CO_2 levels are lowering ocean-water pH. A more acidic ocean affects calcifying organisms, such as corals, with significant effects to reefs, the ecosystems they support, and their ability to protect vulnerable coastlines. The growing body of knowledge about the impact of climate on marine chemistry and ecosystems will also enhance ecosystem-based management efforts.

The ocean has a controlling influence on the path and intensity of major storm systems, such as hurricanes, mid-latitude winter storms, and intraseasonal atmospheric oscillations. Improved understanding of the drivers of these systems resulting in predictive capability will enhance society's ability to prepare and adapt cities and other public infrastructure for the inevitable arrival of severe events and to take advantage of opportunities presented by climate predictions.

Research Priorities

Consistent with the scientific challenges identified in the *Strategic Plan for the U.S. Climate Change Science Program (CCSP)*[23], it is essential to improve understanding of the ocean's role in past, present, and future climate and to educate decision-makers and the public about this role. Phenomena that contribute to climate change and variability include large-scale, long-term coherent variability (e.g., Pacific Decadal Oscillation [PDO], North Atlantic Oscillation [NAO], Atlantic Meridional Overturning Circulation [MOC]); large-scale, nonlinear behavior (e.g., abrupt changes in large-scale ocean circulation); and ocean-atmosphere fluxes (i.e., of carbon, momentum, heat). Applying improved understanding of the ocean's past and present role in climate change and variability will enable better predictions and projections of climate effects on ocean processes and components (e.g., ecosystems). This understanding will help inform current and future management efforts, including coastal land use and development, and alternative energy use and development (e.g., wind and tidal power). A pressing barrier to improved understanding and an item for immediate attention is the lack of sufficient observing capabilities. Determining impacts on the ocean from climate changes, including identifying impacts associated with natural variability and anthropogenic influences, as well as the effect of ocean processes on climate, will be enhanced through coordination of efforts from other societal themes (e.g., factors influencing ecosystem health, natural-resource use, natural-hazard mitigation, impacts to marine operations, and efforts to protect public health).

Research Priority 11: Understand ocean-climate interactions within and across regions.

Ocean regions, such as tropical, polar, and deep sea, can exert powerful influences on, and in turn, be influenced by, climate change. Current understanding of the type or temporal and spatial extent of this influence is limited. Improved short-term climate predictions require a more complete understanding of the influence of global tropical ocean phenomena (as demonstrated by El Niño, PDO, and NAO events and monsoons in the Indian and eastern Pacific Oceans). Increasing global temperatures could lead to an ice-free Arctic Ocean in summer, with potentially widespread impacts, such as changes in polar albedo and ocean-atmosphere heat exchange, alterations in sensitive Arctic ecosystems, and development of new shipping routes (see Enabling Marine Operations). A warmer Arctic can contribute to coastal flooding due to melting ice sheets, and may influence abrupt or longer-term climate change. At the opposite pole, the Southern Ocean is a region of significant biological productivity, CO_2 uptake, and large-scale ocean circulation forcing, with significant climate implications. Finally, the role of the deep ocean must be ascertained, particularly with regards to mitigating climate change (e.g., via carbon sequestration and heat storage). Processes within these regions (e.g., ocean circulation, air-sea interaction, convection, water-mass formation) can have a variety of individual and compounded impacts over many spatial and temporal scales, ranging from regional effects of sea-ice diminishment on Arctic ecosystems to global effects of sea-level rise.

Research Priority 12: Understand the impact of climate variability and change on the biogeochemistry of the ocean and implications for its ecosystems.

Changes in physical properties of the ocean (e.g., heat, freshwater and circulation), as well as biogeochemical properties (e.g., carbon, nitrogen, dust, trace elements, pollutants), can have a variety of impacts, particularly on ecosystems, ranging from coastal watersheds, to shallow and deep-water coral reefs, to open-ocean systems. For example, increasing ocean acidity, altered biogeochemistry, changing current patterns, loss of sea ice, rising sea levels, and modified salinity and sea-surface temperature may irreversibly alter ecosystems. Sustained observations (e.g., global and coastal observatories), process research (e.g., air-sea exchange, ecosystem interactions) and modeling (e.g., coupled models integrating global and regional biological, chemical, and physical data) will help determine fluxes and cycling of biogeochemical variables and help identify and quantify impacts on ecosystems of the greatest importance. These efforts will also help identify subsequent feedbacks regarding the influence of ecosystems on climate (e.g., influence of ocean acidification on carbon uptake and sequestration, desertification, increasing iron dust and subsequent oceanic algal blooms). This effort will serve to support effective management of these ecosystems so that they remain healthy and viable, in conjunction with research examining impacts to and influences on ecosystem health.

Research Priority 13: Apply understanding of the ocean to help project future climate changes and their impacts.

The ocean is getting warmer, more freshwater is being added by melting ice sheets, and more CO_2 is being absorbed from the atmosphere. The ability to predict the timing and magnitude of climate changes due to these and other processes and to understand how such changes will impact society will be enhanced by quantifying the predictability of these processes and increasing temporal and spatial resolution and dynamical complexity of global ocean models. Integrating expanded global and regional ocean observations, coordinated paleoceanographic data and assessments, and enhanced process research into global coupled ocean-ice-atmosphere-land climate models will help determine the past and present influence of ocean processes on climate change, including the potential for rapid or abrupt change (e.g., "tipping points"). Coupled climate models will support improved short-term predictions (e.g., hurricane intensity) and long-term projections (e.g., sea-level rise) that will enable local, tribal, state, regional, and federal policy- and decision-makers to plan for and adapt to effects of climate change.

Necessary Tools

The continuing challenge for climate-related ocean research remains the lack of a robust, integrated system of global ocean observatories (based on quantitative design studies) capable of sustained observations and integrated with international observing efforts. The development of coastal and coastal watershed observing systems is integral to this effort. Observations from the individual components of this system (including *in situ* and remote [e.g., satellite] sensors) must be integrated through data management and communication capabilities that provide open access, searchable content, and routine delivery to all users. In addition to collecting physical, biological, and biogeochemical data, the observing-system effort should include the capability to integrate with future observing systems and incorporate data collected outside of this effort (e.g., sediment and ice-core paleoceanographic data) into a larger database. The databases from these systems should be integrated with the global system. Data systems should have the capability to develop climate-data records for physical and biogeochemical data sets and to reconstruct past ocean states, including the development and refinement of climate proxies.

Ocean and coupled climate model improvements will be necessary to integrate large volumes of observational data and reconstruct past ocean states and predict future states. Increased computational capabilities, including versatile software, efficient algorithms, focused resolution, and capacity, are required to support these advances in data and modeling capabilities.

Sustained and improved satellite and *in situ* sensors are necessary to collect information for a sufficient time period to be able to detect subtle, background climate trends over a broader range of climate parameters, such as currents, salinity, and sea-ice thickness, to draw a more complete picture of fundamental climate processes. Next-generation, *in situ* chemical and biological sensors that collect a variety of information, including data on sentinel organisms and habitats, will also be important. This enhanced data collection must be accomplished while maintaining long-term climate records of key variables.

The nation's coastal ecosys- tem provides a complex array of goods and services, and is influenced by a multitude of climatic and human factors. *Graphic credit: NOAA.*

THEME 5: IMPROVING ECOSYSTEM HEALTH

Multi-faceted and complex, marine ecosystems, including the open ocean, coasts, coastal watersheds, and Great Lakes, provide a wealth of benefits to humankind. They are also finite and vulnerable to overuse or misuse from human activity and impacts from natural events. Comprehensive, well-focused, interdisciplinary research can provide the information needed to balance competing uses of the marine environment, to better predict the impacts of such use, to manage those impacts in a manner that ensures the long-term health[24] and sustainability of marine ecosystems, and to help restore ecosystems damaged from past and current activities or events.

Rationale

Marine ecosystems provide abundant products and services that not only enhance, but also are essential for, life on Earth, such as climate regulation, food and marine products, waste disposal and pollution control, energy sources, and recreation[25]. Marine ecosystems range from the Great Lakes and ice-capped polar expanses to warm tropical seas, from deep-ocean waters and productive continental shelves to coastal waters, including bays, estuaries, and wetlands. They interface with terrestrial ecosystems not only at shorelines, but also through watersheds and the atmosphere. Marine ecosystems are also subject to individual and cumulative impacts from an array of human activities (e.g., resource extraction, shoreline alteration, modified land use, increased flux of compounds, introduction of invasive species) and natural events and processes (e.g., hurricanes).

Much remains to be learned about the structure, function, and vulnerability of ecosystems, including factors controlling ecosystem stability and function; processes across the air-land-water interface, including the effects of human activities; linkages between ecosystem types (e.g., riverine and coastal, continental shelf and deep pelagic) and trophic levels; and the role of marine ecosystems in the ocean-Earth-atmosphere system that enables life in its myriad forms. Scientific understanding provides the foundation for effective management and governance that will help ensure the sustained vitality and diversity of marine ecosystems, as well as the social, economic, ecological, and human- health benefits society derives from them.

Research Priorities

Effective and adaptive ecosystem-based management will require a more complete understanding of the mechanisms governing ecosystems and an ability to accurately predict society's effects on them. Expanding understanding of ecosystem structure, function, complexity, and stability will require investments in exploration, inventory, and novel methods to investigate ecosystem components and their interactions. Such investigations will be complemented by research in other societal themes, all of which are linked to marine ecosystem health. This improved understanding and predictive capability will enable effective evaluation of the success of management efforts (e.g., MPAs).

Research Priority 14: Understand and predict the impact of natural and anthropogen ic processes on ecosystems.

The complex relationships that determine ecosystem structure and function vary spatially (e.g., local, regional, basin-wide, global) and temporally (e.g., seasonal, annual, decadal, centennial), and incorporate various feedback effects. Shifts in ecosystem structure can be induced by climate-ecosystem interactions (e.g., impacts from El Niño/La Niña, increases in ocean temperature, ocean acidification), human activities (e.g., watershed activities, sediment/nutrient/contaminant flux, resource extraction), and productivity- driven or predation-driven trophic mechanisms (e.g., invasive species). Understanding these relationships requires incorporating new and existing knowledge and observations at appropriate temporal and spatial scales to explore the full range of physical, chemical, biological, and ecological mechanisms that determine observed ecosystem structure, function, and variation. Improved ecosystem forecasting requires: (1) comprehensive analyses of natural and anthropogenic changes in physical, biological, and chemical properties and their individual and cumulative impacts on productivity and overall ecosystem health; (2) assessment of dispersal mechanisms for marine organisms; (3) development of next-generation trophic dynamics models, spanning multiple trophic levels; and (4) process studies and model development to assess impact (loss) and recovery responses to natural and anthropogenic stressors.

Research Priority 15: Apply understanding of natural and anthropogenic processes to develop socioeconomic assessments and models to evaluate the impact of multiple human uses on ecosystems.

Understanding human impacts on marine ecosystems, whether positive (restoration) or negative (degradation), will require integrating traditional ocean science with socioeconomic science. To determine and predict society's total impact on marine ecosystems, social and economic factors (e.g., energy uses, coastal and watershed development, land use, water use, resource-use perception) that determine how society views and uses marine ecosystems must be assessed and modeled. Existing non-market values of ecosystem services and processes are often geographically and topically fragmented, rendering temporal and spatial analysis of non-market valuation difficult. Research efforts must include developing new methods to evaluate non-consumptive use of ecosystem services and characterizing the value society places on competing uses. To ensure sustainability of ecosystem goods and services, those methods must consider the rights of future generations and include discounting procedures for adjusting cost-benefit analyses over time.

Research Priority 16: Apply understanding of marine ecosystems to develop appropriate indicators and metrics for sustainable use and effective management.

A robust suite of indicators of ecosystem structure, function, productivity, and services must be evaluated and implemented at multiple scales (local, regional, basin-wide). These indicators will help assess factors that stress and degrade ecosystems (e.g., eutrophication and harmful algal blooms, loss of coastal wetlands, shoreline development, overuse of harvested species, invasive species, introduction and cycling of contaminants, changes in biodiversity, ecosystem productivity or resilience). Additionally, indicators and metrics are needed to help monitor the restoration and recovery of degraded ecosystems. Input from a range of scientific disciplines is needed to identify such indicators or develop them where existing indicators are insufficient or gaps are identified. Metrics and indicators must evolve based on new scientific understanding and management needs and goals. Such metrics and indicators, evaluated through integrated ecosystem assessments, will provide feedback for assessing management efficacy and a basis for incorporating new information and understanding to adapt and improve management practices.

Necessary Tools

These ambitious but essential research priorities depend on the continued development and implementation of ocean-observing capabilities for assessing physical, chemical, and biological properties of marine ecosystems at appropriate temporal and spatial scales, as well as for assessing human impact on key ecosystem properties, such as productivity, diversity, and resilience. Effective observing capabilities must provide essential data on ecosystem types, ranging from terrestrial watersheds, to productive coastal and continental shelf regions, to the deep, pelagic realms. The information gathered will support essential ecosystem-related analyses and provide necessary input into effective, reliable ecosystem models. Collection of such data will require extensive infrastructure, including research vessels, automated buoys, and autonomous vehicles for short- and long-term sampling of water-column properties;

satellite-based assessment of surface characteristics (e.g., temperature, biogeochemical properties, surface currents, wave heights); *in situ* observatories in the ocean, on the seafloor, and across the land-water interface; shore-based facilities for sample analysis, experimental manipulation, and observing-system maintenance; and a range of survey (e.g., mapping) capabilities. Improvements in information technology and infrastructure also will be essential to ensure that data assimilation, analysis, and modeling tools are available accommodate and enable the integration of ecosystem and socio-economic data.

Ensuring an effective and adaptive ecosystem-based research and management approach requires an investment not only in technology and infrastructure, but also in the education systems that produce the scientists and managers needed to implement this approach. Research efforts also require a diverse workforce knowledgeable in natural aquatic and terrestrial sciences (e.g., biogeochemistry, taxonomy, systematics[26]) and social sciences (e.g., sociology, economics), as well as individuals trained in transforming data into information products for end users.

Theme 6: Enhancing Human Health

The ocean can be a source of health hazards. Understanding the causes of health hazards and how they can be mitigated or managed will lead to fewer illnesses from contaminated seafood, polluted waters, known and emerging disease-causing microbes, and harmful algal blooms (HABs). In the next decade, efforts to enhance human health will focus on the cause, prevention, and treatment of disease. The ocean also holds abundant resources that convey a variety of health benefits to humans. Exploration of new habitats, combined with emerging biochemical and biotechnical techniques, will promote discovery and development of bioproducts that promote human health.

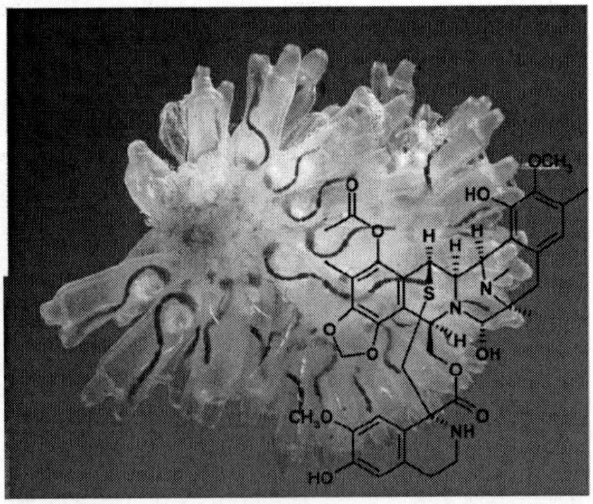

Yondelis® (trabectedin, ET-743) was originally derived from the mangrove tunicate *Ecteinascidia turbinata*, which occurs through-out subtropical mangrove and sand-flat habitats in Florida, the caribbean, and Mediterranean. it is now produced syntheti- cally and is being developed by PharmaMar in partnership with Johnson & Johnson Pharmaceutical Research & Development (J&JPRD) for the treatment of soft-tissue sarcomas and ovarian3 cancer. *Photo credit: Harbor Branch Oceanographic Institution.*

Microsystis bloom, August 2003 as seen in Landsat 7 photo of western Lake Erie. From Backer, L.C. and Dj. McGillicuddy. 2006. Oceanography 19(2), Figure 8 by Ohio View/T. Bridgeman.

Rationale

Waters from the open ocean, coasts, coastal watersheds, and Great Lakes present health risks to people through the consumption of contaminated seafood or drinking water (containing pathogens, toxins, or current or emerging contaminants); direct contact through recreation, work, and weather events; and indirect contact such as breath-ing salt air that may contain algal toxin aerosols. Illnesses due to the consumption of contaminated shellfish and finfish are estimated to affect hundreds to thousands of people per year[27]. However, these figures are uncertain due to major gaps that exist in illness reporting and epidemiological knowledge of seafood-caused[28] human illnesses[29]. Given the number and diversity of marine organisms and associated pathogens, the ocean should not be overlooked as a potential reservoir of pathogenic threats to humans[30]. Isolating the causes and impacts of these hazards will help protect human health and safeguard the quality of the seafood supply.

In addition to providing a wealth of known resources such as food, the ocean is full of potential human-health assets. In recent decades, scientists have discovered whole new ecological communities in the ocean with unique biochemical systems, such as those associated with thermal vents and hydrocarbon seeps. These communities, along with ones yet to be discovered, hold huge possibilities for development of products to improve human health and well-being. Prior discoveries that now have practical applications include pharmaceuticals (i.e., anti-cancer drugs), diagnostics (i.e., endotoxin detection), molecular probes (used in biochemical process/disease research), and nutrients (e.g., xanthophyll feed additives derived from algae)[31].

Research Priorities

The ocean community faces many challenges in addressing the relationship between the ocean and human health. Adequate process studies and assessments, monitoring, and data sharing and interpretation are needed to determine sources of human health risks, assess and predict risks to human health from those sources, and support the needs of a wide variety of end users. Expanded exploration and investigation of the ocean environment will contribute to our current understanding of the potential benefits from the ocean. These efforts also strongly interface with initiatives to improve ecosystem health and understand ecosystem processes, impacts of climate change, and effects of natural hazards.

Research Priority 17: Understand sources and processes contributing to ocean-related risks to human health.

Investigations into ocean-related risks should target ocean ecosystems and processes that impact human health. The introduction (e.g., runoff, atmospheric deposition), cycling (sediment, water column, atmospheric), and effect (chronic and acute) of current and emerging pathogens, toxins, and contaminants (e.g., mercury, flame retardants, endocrine disruptors, hydrocarbons) must be evaluated in coastal, ocean, and Great Lakes environments. Assessments of food webs can help determine the fate of these compounds (e.g., bioaccumulation, biotransformation, biomagnification), which can impact human health. Coastal and watershed processes and mechanisms that stimulate and sustain HABs must also be ascertained. Given current climate trends, the influence of climate change and variability on water-borne diseases (e.g., cholera) must be determined. Identifying, developing, and implementing new and improved sensors, rapid assays, models, and methods will enable assessment, monitoring and prediction of a variety of human health risks, such as HAB onset, extent, and duration; infectious disease potential (including microbes that can cross from animal to humans); and pollutant source and loading. These new sensors and methods will also provide crucial information related to the health of marine wildlife populations. The resulting improved understanding will enable development of mitigation strategies and technologies, such as stormwater-runoff and wastewater-discharge controls, and new approaches to minimize health-related impacts of hurricanes and other natural hazards.

Research Priority 18: Understand human health risks associated with the ocean and the potential benefits of ocean resources to human health.

While many ocean-based benefits and risks to human health are known, most are poorly understood and their direct health effects on humans are inadequately documented[32]. Benefits to humans associated with consumption of seafood (e.g., improved cardiovascular health, cognition) should be better characterized. Investigations should also quantify risks and impacts of exposure to health hazards (e.g., contaminants, pathogens, and toxins) via various pathways, including skin contact, respiration (e.g., inhalation of airborne materials), and consumption of food or water and determine the incidence and severity of human illnesses. Efforts to quantify risks and benefits require improving assessment methodologies and enhancing the accuracy and sophistication of epidemiological studies. Initial epidemiological studies focusing on risks should be directed towards those human populations expected to be most at risk (e.g., children, pregnant women, individuals with compromised respiratory or immune systems, people who spend a large amount of time on or in the ocean, coasts and Great Lakes—such as commercial, recreational, or subsistence fishers) and on diseases in ocean species that may serve as sentinels for new or ongoing threats. Ocean data and modeling should be integrated with assessments and epidemiologic studies to define exposures and focus risk assessments.

Research Priority 19: Understand how human use and valuation of ocean resources can be affected by ocean-borne human health threats and how human activities can influence these threats.

Coastal and Great Lakes environments are highly attractive places for recreation and they play major roles the nation's culture and economic vitality. However, the use and value of these environments and their resources may be significantly impacted by ocean-borne threats. These impacts include, but are not limited to, beach closures, constrained resource use (e.g., fishery and shellfish harvest area closures), and reduced marine operations and recreation due to contamination by pathogens, toxins, or pollutants, and other humanhealth threats (real and perceived). Human activities, such as coastal development, contribute to the onset and persistence of many of these human health threats. Understanding and predicting the relationship between social and economic drivers (e.g., shoreline development, tourism, recreational and subsistence fishing) and human-health threats will require integrating socioeconomic data and investigations with ecosystem-based studies of health threats at appropriate temporal and spatial scales. The integration of information from natural and social sciences will, in turn, help support management and mitigation efforts.

Research Priority 20: Apply understanding of ocean ecosystems and biodiversity to develop products and biological models to enhance human well-being.

The intricacies and biodiversity of ocean ecosystems provide unparalleled opportunities for discovery and development of useful materials. Collaborative research efforts incorporating multiple disciplines (e.g., evolution, ecology, genomics, pharmacology) should focus on expanded exploration (conducted with care to prevent transfer or introduction of wildlife diseases or invasive species), assessment (e.g., genomics, advanced non-culture based methods), and development (e.g., biosynthesis) of ocean bioproducts (e.g., pharmaceuticals, nutrients, diagnostic tools, reagents, enzymes). Research efforts should also include developing the capability to use marine species as mechanistic models for the study of diseases, toxicology, and biochemical processes relevant to human health; and identifying and using appropriate sentinel species (e.g., aquatic, avian) and habitats that may serve as early-warning systems of potential ocean risks to humans.

Necessary Tools

A system is needed to record illness data due to pathogens, biotoxins, and chemical pollutants. Integral to this system is expanded environmental monitoring and infrastructure. Such a system would require collection of human and animal health data and relevant environmental data (via remote, moored, and mobile platforms and sensors as part of an integrated ocean-observing system), coupled with new data integration and interpretation capabilities, and rapid data-communication plans. Once in place, the system could serve to monitor and predict ocean conditions that place people and other animals at risk or that may be favorable to human health.

Sensors capable of detecting biological and chemical parameters, such as microbial species and densities, and toxin and contaminant identifications and concentrations must be enhanced or developed. Remote sensing of ecological changes and real-time, high-frequency (temporal and spatial), *in situ* monitoring combined with modeling efforts (statistical-empirical and mechanistic) are necessary to inform understanding of ecosystems and to track and predict outbreaks and other impacts. Designing advanced sensors for hostile and extreme environments will enable more extensive ocean explora-tion. Developing, enhancing, and applying new methods and tools at shore-based marine laboratories and other facilities with specialized instrumentation (e.g., for large-scale gene sequencing) and computational resources in areas such as genomics, proteomics, and bioinformatics will expand surveying and screening capabilities. Improved coordination of federal and state data systems is needed for toxic algal-bloom monitoring, pathogen-source tracking, marine-disease surveillance, and medical-illness reporting, and in the design and development of a national database of monitoring and screen-ing methodologies.

Many of the research efforts require interaction between oceanography and other disciplines, such as biomedicine and public health, to provide a broader understanding of the issues and to communicate the results of this research to support effective decision-making.

Opportunities for Progress

Developing the Tools

To address key ocean-related societal issues and to apply existing understanding to support meaningful decision-making requires development and implementation of a broad infrastructure. Each societal theme outlines the tools needed to address the respective research priorities, such as vessel time and new ship-based technologies (Stewardship of Natural and Cultural Ocean Resources, Improving Ecosystem Health), shore-based facilities with specialized instrumentation (Enhancing Human Health), and reactive deployment capabilities (Increasing Resilience to Natural Hazards), all of which are critical to pursing the research priorities. Several key infrastructure needs were common across many themes and are described below.

Observing Systems

The need to acquire, manipulate, analyze, and deliver requisite information about the ocean calls for development of an integrated ocean-observing system that includes coupled observational and research components. This system will consist of existing capabilities (e.g., established *in situ* networks, current and planned satellite missions) that will be continually integrated with future systems (e.g., seafloor networks, regional observing systems). Covering a diverse suite of spatial (coastal watersheds to polar seas) and temporal (microseconds to decades) regimes, an integrated ocean-observing system will provide data (in real time or near real time) to address a broad spectrum of societal and ecological issues, including human-dimension processes; build the foundation for advancing basic understanding of open ocean, coastal, coastal watershed, and Great Lakes processes; and revolutionize scientific and public access to the ocean. An integrated system represents an ambitious effort of the national and global ocean community to establish and maintain a robust, adaptive, continuous presence in the ocean and provide critical information and models to users and stakeholders. Given the interconnected nature of land-sea-air systems, an integrated ocean-observing system should be capable of interfacing with current and future terrestrial, aquatic, and atmospheric observing systems and be closely coordinated with the Global Ocean Observing System, as part of the Global Earth Observing System of Systems (GEOSS)[33]. Several key steps must be taken to fully implement the system:

- Fully integrate existing (e.g., buoy networks) and future (e.g., surface-water-quality monitoring networks) components of a national observing system ("national backbone") to provide nationwide coverage of the most commonly required ocean parameters.
- Expand regional and local coastal observing-system capabilities (fixed and mobile) to provide greater density and diversity of observations on a regional basis, and address region-specific observational needs, such as watershed-specific monitoring and land- water fluxes.
- Maintain continuity of existing satellite missions and incorporate new remote-sensing capabilities into future missions. Integral to this effort is the need for continued sensor development, validation, and calibration that will ensure delivery of functional data to users.
- Incorporate new and existing biological, chemical, physical, and geological measurements into the observing system, and expand sensor development for biological, chemical, and physical parameters.
- Implement data-management and communications plans that will modernize existing data systems. Integrate current and future systems by promoting interoperability among system components; developing metadata standards; integrating new components; and enabling data discovery, availability, use, and archiving of the data streams collected by the observing system and other observing assets (e.g., vessels).
- Integrate the data needs for decision-support tools into the development of observational and information systems and models.

Models

Combining existing data sets with scientific and social theory and modeling will enhance understanding and address user needs by providing a window on past conditions, permitting an assessment of present conditions, and enabling predictions of future conditions. Models may be developed for several reasons, including *exploring* the relationships among physical, chemical, biological, and socioeconomic variables; *hindcasting* previous conditions in an area; and *forecasting* the impacts of environmental change or management practices.

Models can comprise a certain configuration, including parameters, domain, and boundary conditions. The term model can also be used to describe a modeling system, which incorporates an array of algorithms (a set of sequenced, computational instructions) from which the most appropriate combination can be selected for a particular application. The discussion of models throughout the societal themes incorporates both model configurations as well as modeling systems.

Key elements to enable modeling efforts nationwide:

- Expand translation of data and insight derived from ocean research and observations to improve representation of this information in models.
- Develop new application-specific model configurations where necessary.
- Consolidate existing modeling capabilities into a flexible and comprehensive unified ocean-modeling software environment, building on the diverse existing community and developmental ocean models.
- Improve model analysis and capabilities through model diagnostics and intercomparison activities.
- Conduct systematic best-practice studies to develop guidance in the selection of optimal algorithms and techniques for particular modeling applications.
- Expand development of high-resolution, global ocean model configurations for use in real-time, short-term forecasts out to days or months, reanalysis of historical ocean state, and multi-century climate forecasts.
- Expand current computational capacity to assimilate increasing amounts of data provided by ocean-observing assets and to support and process model advancements incorporating increased process and mechanistic understanding.

To ensure the effective use of model output, developers and users must address several issues, including how the temporal and spatial scales of the input data compare to the outputs and products, how uncertainty in data sets and model calculations is propagated, and how the models are intended to be and are actually used.

Making a Difference

Information to Support Decision-Making

Scientific results must be actively translated into readily understandable information and made available to the public, and their trustees and elected representatives. Information from managers and decision-makers will, in turn, help develop and inform research efforts.

Develop mechanisms to further integrate research results into adaptive management efforts

Effective resource management must include mechanisms to design, evaluate, and modify management approaches. Science-based management decisions require a close coupling of problem formulation and data collection. A two-way interface connecting research results and management efforts will support more rigorous adaptive management, where management efforts are designed and can be modified based on empirical evidence of system response. This interface between researchers and decision-makers, enabled through extension programs and technical training, will allow managers to acquire the information they need, and researchers to help focus research to priority areas identified by resource managers.

Establish dedicated mechanisms to translate research results into readily used products

To enhance communication among sectors (e.g., researchers and resource managers) and enable translation of results into readily used information products, communication strategies must be investigated, developed, and enacted. Part of these communication efforts may include the development of professional communicators to assist researchers and end users in developing more accessible, derivative products. These professional communicators would be cognizant of both scientific objectives and management goals. Communicators can also help groups exchange information, including research needs and research results, and enhance communication among disparate sectors.

Foster mechanisms to transition developing technologies into operational capabilities

Many of the research priorities outlined in this document require new sensors, platforms, data-management technologies, and models. However, there is often a gap between the development of these capabilities and their use as robust components of integrated systems. Fostering communication among agencies, researchers, engineers, and end users will support efforts to overcome technical barriers and promote effective development, implementation, and use of these technologies.

Establishing an Ocean-Literate Nation

The goal of ocean education is a nation whose citizens are good stewards of the ocean and who possess the knowledge to make informed decisions about their interactions with it. This goal can only be attained through improved education efforts for the entire spectrum of ages ("K–gray") through formal and informal education—to ensure ocean literacy for the general public, development of a well-trained workforce, and more effective communication.

Ocean research provides the information needed to develop education, communication, and management materials that are based on sound science. Ocean-research activities also provide a training ground for a range of ocean professionals, an aspect of the research enterprise that is particularly important as ocean science incorporates other fields, such as medicine and social sciences. Efforts to enhance ocean literacy will support and, in turn, be supported by efforts to promote fundamental math and science skills. Coordinated and sustained education and training efforts for diverse groups, from government agencies and the academic community, to school systems and scientific societies, will promote a knowledgeable, ocean-literate public and a larger and well-prepared workforce. These efforts will help retain U.S leadership in science and technology.

Expand and sustain formal and informal education efforts

Currently, ocean science is rarely included in formal education settings. Assessing current ocean-based curricula and then developing well-focused, ocean-related materials will help incorporate ocean science into K–12 curricula, in coordination with ongoing national science education efforts. Ocean science is an ideal mechanism to convey fundamental scientific concepts and mathematical skills. Providing professional-development opportunities for teachers ("educating the educators") and expanding and acknowledging the role of scientists, managers, and policy-makers in education and outreach efforts are additional key steps in expanding ocean science in formal education settings.

Incorporation of education goals will add value to research efforts. Integrating ocean-observing data and information product into K–12 classrooms will help realize the full potential of the observing systems and provide real-world application to the learning environment. Strategies for using these data need to be coordinated at a national level.

Aquaria, museums, marine laboratories, science centers, zoos, and other informal-education centers welcome over 142 million visitors a year and play a major role in educating the public about the ocean[34]. Media outlets, such as newscasts, television specials, movies, magazines, and online portals, provide necessary information, such as hazard warnings, and provide a window into the world of ocean-science exploration and discovery. Citizen science initiatives can also serve to engage citizens with the ocean while promoting ocean research and monitoring efforts. Expanding and supporting these forums for lifelong learning will enable the transfer of understanding about the ocean from the research community to the general public.

Advancements in formal and informal education must be accompanied by efforts to address and capitalize on differences in culture, geography, language, and values to reach a diverse audience. These efforts should include strategies to recruit and retain students from all population segments.

Expand intellectual capacity

The diversity of technical and research requirements presented in this document illustrates the need for a well-trained workforce, including individuals with disciplinary (e.g., biology, chemistry, geology, physics), interdisciplinary (e.g., biogeochemistry, aquatic, and terrestrial sciences), and inter-community (i.e., connecting oceanography, biomedicine, engineering, economics, information technology, public health, community planning) training. This need is particularly evident as ocean science expands from having single-discipline foci to addressing topics that transcend natural sciences and incorporate social and economic sciences. The workforce necessary to use the research results outlined in this document must include individuals who are well versed in scientific discovery and understanding, and who possess the means to effectively communicate research results to inform management decisions and policy development. Developing this workforce requires new models by which students are trained and evaluated that recognize the value of interdisciplinary work. Individuals trained in research must also be educated in policy, regulation, and resource management to better understand and address user needs, where appropriate.

Near-Term Priorities

The Joint Subcommittee on Ocean Science and Technology (JSOST) developed four near-term (2–5 years) priorities, each with equal weight, to focus initial research efforts. However, these efforts do not preclude other activities towards all 20, longer-term (7–10 years) research priorities. These near-term priorities, which reflect multiple aspects of the national research priorities and the critical tools needed to address them, were selected from a larger suite of efforts using the criteria outlined in Identifying Ocean Priorities, with an added focus on impact (i.e., the value of the work), urgency (i.e., the need for a concentrated effort over the next two to five years), and partnerships (i.e., the effort will maximize collaborations among agencies and external partnerships). The primary societal themes from which the near-term priorities were developed are highlighted in the individual discussions.

1. Forecasting the Response of Coastal Ecosystems to Persistent Forcing[35] and Extreme Events

Coastal ecosystems are subject to a variety of forcings, ranging from extreme events (Increasing Resilience to Natural Hazards), to human activities (Improving Ecosystem Health), to changing ocean conditions (The Ocean's Role in Climate). Understanding the response of natural and constructed landscapes and ecosystems (e.g., algal blooms, hypoxia, coral-reef bleaching, decline in sea ice) to these forcings; forecasting the frequency, intensity, and impact of those forcings; and providing tools to develop policy and management responses are integral to constructing more resilient structures and communities, and protecting the natural environment. Research and observations will focus on establishing the basis for both short-term forecasts and long-term, probabilistic assessments of coastal vulnerability to extreme events, persistent natural processes, and human influences across the coastal zone. This effort will enhance regional observing systems and models, integrate substantial existing observations, and incorporate new observations to address critical regional data gaps. Data and information products will be made widely available to diverse end users through a national ocean-observing capability. Results from this effort will inform hazard mitigation and response plans, provide forecasting data to support navigation safety, and assist regional resource managers and public health officials in sustaining ecosystem and public health, and promoting hazard resilience.

Observations of physical characteristics and processes, including material inputs from adjacent watersheds and contributing airsheds, ocean influences on hurricane intensification, and characterization of submerged and coastal landscapes, will be integrated to support data-assimilative modeling of, for example, water quality, nutrient, sediment, and contaminant transport; waves and water levels; and the coastal response to hurricane processes. Biological observations, including new DNA-based techniques, will enable development of coupled physical and biological models of ecosystem-level response to various stressors. Coupled research, observations, and model development might focus on identification, quantification, and transport of pathogenic microbes and various species of harmful algae, and lead to robust, timely forecasts of human-health threats and natural-resource impacts. Additionally, linking environmental-quality data with public-health surveillance activities will support modeling and prediction of the geographic expansion of potential health risks from hurricanes and specific waterborne vectors, toxins, and pathogens.

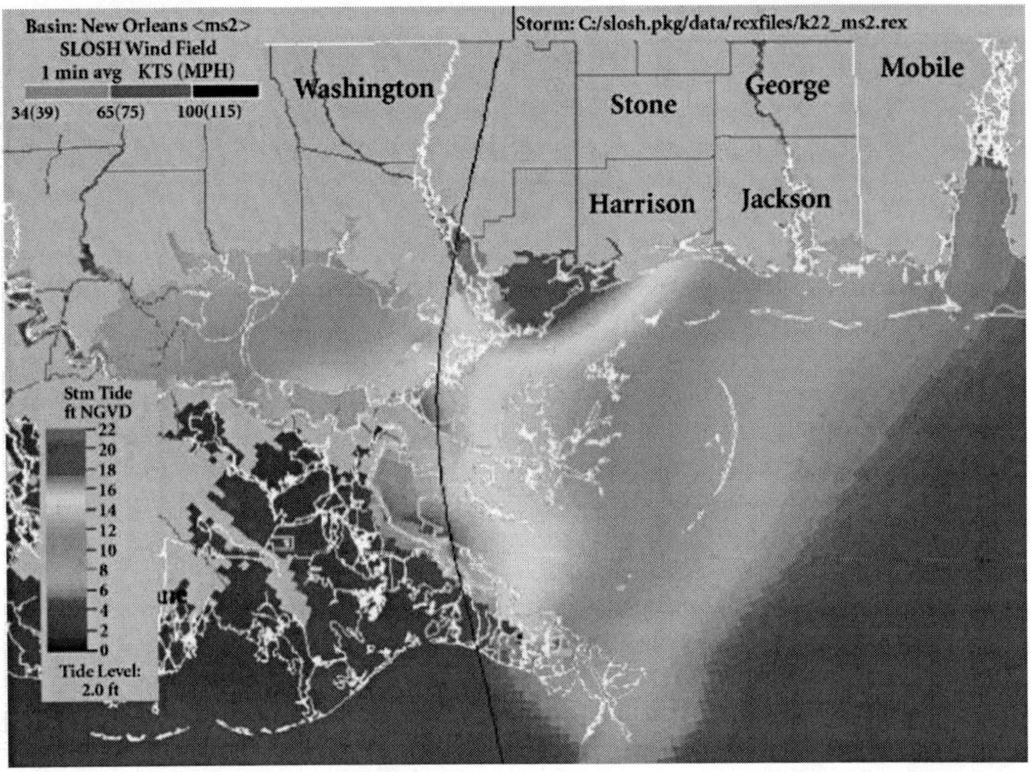

Storm surge data posted by the National Hurricane center at 9:20 AM cDT 8/28/05 provided critical data on the potential impacts of Hurricane Katrina. *Graphic credit: NOAA.*

This effort will build upon extensive existing data sets, surveillance, observational, and modeling capabilities; promote the transition of models from research tools to operational applications; and support the establishment and linkage of regional and national oceanwater-quality data networks. Decision-support models will inform prevention strategies, rescue and recovery operations, spill tracking, safe maritime navigation, water-quality forecasting, and resource assessment and management, taking into account the vulnerability of ecosystems as well as their capacity for mitigating harmful impacts. Focus areas for implementation will be identified to build upon existing assets where integration and enhancement provide the greatest opportunity to impact public safety and public, economic, and environmental health.

2. Comparative Analysis of Marine Ecosystem Organization

Forecasting marine-ecosystem responses to management strategies requires an understanding of the complex dynamics that control and regulate ecosystem processes (Improving Ecosystem Health, Stewardship of Natural and Cultural Ocean Resources). Management of marine ecosystems can be improved by elucidating the underlying dynamics of these systems at a variety of scales. This effort will provide a greater basic understanding of ecosystem processes and practical tools for evaluating the effectiveness of local and regional adaptive ecosystem-based management efforts.

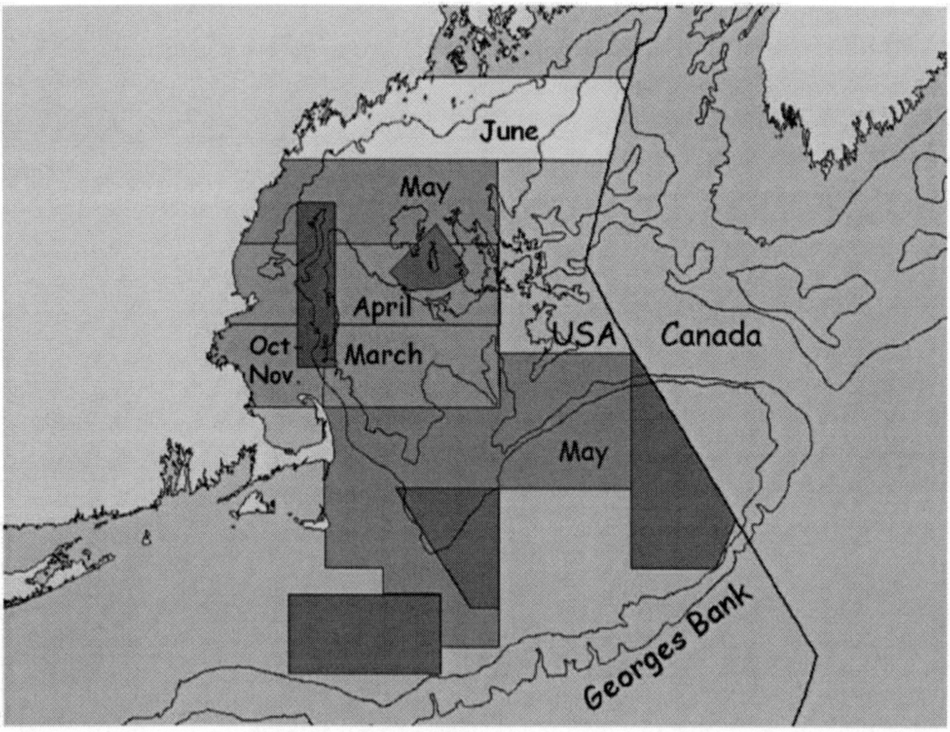

Marine Protected Areas (MPAs) off the northeast united states. dark blue areas are closed to certain fishing techniques year-round, while other regions are closed in certain months. *Graphic credit: NOAA.*

Ecosystem-based approaches emphasize interactions among components and the impacts that various human activities have on productivity and organization. Forecasting these impacts requires understanding complex dynamics controlling: (1) productivity of various trophic levels, (2) predator-prey interactions, (3) connectedness of sub- populations, (4) impacts of natural climate variation, and (5) anthropogenic pressures. Because classical controlled experimentation provides limited information on the complex dynamics of marine ecosystems, two types of analyses will be undertaken. The first approach will construct and apply various classes of energy-budget and dynamic models to managed marine ecosystems to enable greater understanding of the impacts of human activities by contrasting biomass changes according to trophic level. The second approach will compare systems where marine managed areas have been enacted to conserve species and ecosystems[36]. Such comparisons will include before and after area designation contrasts where sufficient data are available, and inside versus outside comparisons for established managed ecosystems. Mapping efforts will include ecosystem-scale characterization, design of interpretive products, and provision of tools to assimilate and disseminate geospatial information in support of research, observations, modeling, forecasting and management decision support.

Candidate ecosystem types for inclusion in this study may include the sub-Arctic, continental shelves, coral reefs, and estuaries. Analyses of these ecosystems will focus on how feedbacks influence ecosystem productivity, biodiversity, and conservation of managed species through comparisons using consistent modeling frameworks. Evaluation of the effects of management efforts will involve assimilation and synthesis of existing biological information by trophic level, linkages to higher levels, and impacts on human-use patterns (e.g., displacement of human activities from marine protected areas and their socioeconomic effects). These efforts will help ensure that effective ecosystem-based management strategies are based on sound scientific understanding.

3. Sensors for Marine Ecosystems

Advances in ocean sciences have been made possible by technological innovation. The development of new sensor capabilities, integral to many of the research priori-ties, will help realize the full potential of *in situ* observing networks and satellite-based observations (Developing the Tools) and enhance understanding of marine ecosystems (Improving Ecosystem Health, Enhancing Human Health). Currently, significant limitations exist in observational capabilities and associated methodologies. Advancements in sensor capabilities can revolutionize understanding of the ocean environment by providing information at temporal and spatial scales not currently available.

Immediate improvements can be made to *in situ* instrumentation in eco-genomics, to interdisciplinary ocean observing, and to sensors in support of satellite-derived ocean color (used to estimate biological [phytoplankton] and biogeochemical [carbon] materials in surface waters) and new biological and biogeochemical observations. Creation of a common library of genetic "barcodes" is a critical step in the development of fast, economical genetic screening procedures to identify marine organisms. Such procedures will help reveal basic ocean processes controlling biodiversity and productivity, such as the distribution and abundance of harmful algae and pathogens. Advancing novel capabilities and moving high-impact/high-utility sensors from research to broad operational use will allow for physical sensors capable of providing a better understanding of ocean transport and fluid velocity, and chemical sensors that can reliably detect toxins and ecologically important nutrients. Enhancing in-water instrumentation used in ground-truth exercises will improve the quality and usefulness of ocean-color data provided by satellites, and will aid in the development of new space-based, ocean biological and biogeochemical measurements. Improving ground-truth and *in situ* observations will significantly advance understanding of—and ability to model—ocean systems (ecology, biology, carbon) and their role in the Earth system, as well as the potential impact of human activities. Parallel advances in power-supply technologies and data transfer must also occur to ensure *in situ* sensor operation in a variety of environments and within a variety of networks. Advancing sensor capabilities will enable new, multiscale observations that provide information needed to better define marine resource management options, help understand processes that influence ecosystem health, serve as the basis for forecasting ocean-related risks to human health and safety, and shed light on the impact of climate variability and change on the ocean, marine life, and humans. Expanding sensor capabilities will build on existing mechanisms of support (e.g., agency programmatic priorities) and coordination (e.g., public-private partnerships), and development efforts (e.g., laboratory-based prototypes, expanded capability of existing satellite sensors).

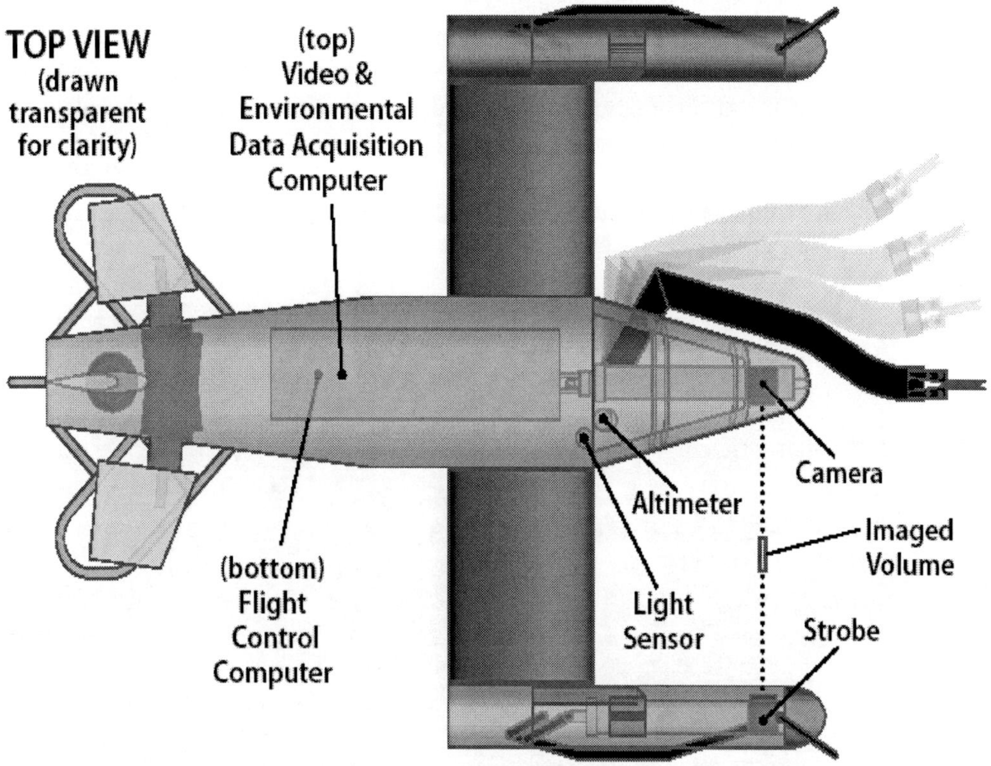

Video plankton recorders (VPR) and other advanced biosensors can be used to provide rapid and synoptic data on biological phe-nomena (such as copepod abundance on Georges bank). *Graphic credit: Jayne Doucette, Woods Hole Oceanographic Institution.*

4. Assessing Meridional Overturning Circulation Variability: Implications for Rapid Climate Change

The ocean plays a critical role in global climate. Thus, incorporating ocean observations and understanding into an integrated, Earth-system-analysis capability is needed to assess the current state of the climate system. Specifically, the Atlantic Ocean MOC, an element of the global-scale ocean circulation responsible for long-term climate variations, hasalso been identified as a key process related to rapid or even abrupt climate change (i.e., changes over a few years to a few decades) (The Ocean's Role in Climate). Rapid changes the MOC could have a profound effect on European and global weather patterns and ecosystems in the Atlantic Ocean.

Assessing the potential for future abrupt climate changes and developing the capability to predict their occurrence will require a national program that incorporates: (1) ocean observations, including *in situ* instruments (e.g., currents, temperature, carbon), satellites (e.g., sea surface height, surface vector winds), and ocean data computational and assimilation capabilities to provide routine, basin-scale analyses; (2) now-casting (an assessment of current conditions); (3) model development for decadal forecasting; (4) past climate- change reconstructions; and (5) climate-impact assessments. Ongoing national and international observing efforts and process studies aimed at improving climate and ocean general circulation models provide the foundation upon which to expand the understanding, observation, and prediction of the MOC and its impacts. New research on the MOC will build on the legacy of seasonal-to-interannual climate prediction systems developed since the 1980s based on Pacific and Atlantic tropical ocean variability. Establishing the basis for a long-term monitoring system for the MOC over the next two to five years will provide the observational data needed to challenge and improve climate models, and to more accurately establish the true variability of the MOC and its effects.

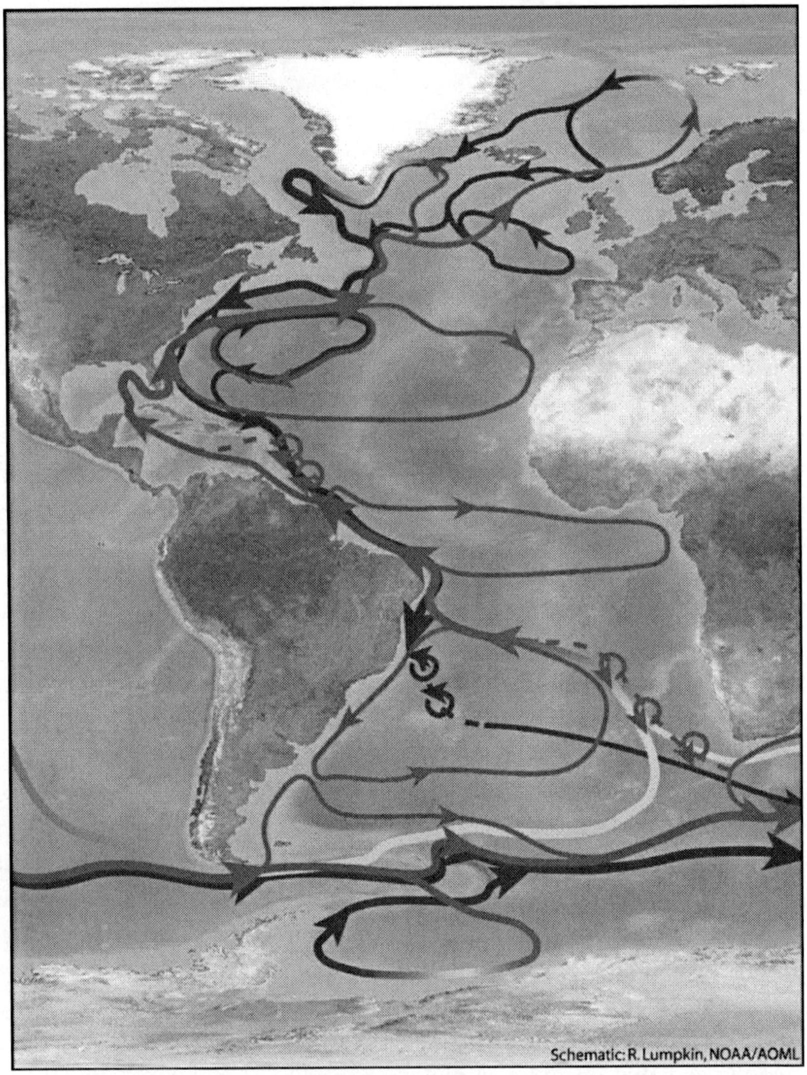

Pathways of the overturning circulation in the Atlantic Ocean that represent the large-scale conversion of surface waters (red arrows) to deep waters (blue arrows) that have global impacts. The Atlantic overturning circulation has many complex and interacting parts (e.g. rings, interbasin exchanges), but the most important aspect is the large heat it carries and its apparent sensitivity to the hydrological cycle and climate change. *Graphic credit: NOAA.*

IMPLEMENTATION STRATEGY

Addressing the national ocean research priorities requires a nationwide effort. The involvement of end users of scientific information, including resource managers, public policy-makers, and individual citizens, will enhance the impact and value of research efforts. Given the significant investment state and local entities make in collecting and disseminating environmental data and supporting research, and the emerging role of private foundations in supporting ocean science, federal efforts represent a key, but not exclusive, element of the national research effort. A well-planned and robust federal effort will provide leadership, focused initiatives, and resources against which non-federal efforts can be better leveraged.

The implementation strategy described in this section is a set of operational principles and conceptual guidelines developed by the federal agencies (the JSOST) to address the national ocean research priorities. This section lays out the characteristics of implementation, the roles of various sectors, mechanisms for collaboration, the need for an infrastructure assessment, mechanisms for research translation, strategies for assessment and evaluation, and mechanisms for budget and plan updates. No one group or sector, including the federal agencies, is expected to address the priorities (near- and long-term) alone; thus, the efforts will reflect differing levels of investment from different sectors, depending on agency missions and sector interests and needs.

This section provides direction to ensure the ocean research priorities are successfully addressed; as such, it outlines an implementation strategy, not a detailed implementation plan. This section does not stipulate specific federal-agency actions or budgets, provide timelines for activities, or mandate how other sectors should respond to the research priorities independent of federal collaboration. Those activities—by virtue of their dynamic nature—are best defined in specific program and project plans. This strategy also leaves open the possibility that dialogue with non-federal partners will play a role in shaping detailed implementation plans intended to address specific research priorities. Applying this implementation strategy will lay the foundation for the multi-tier and multi-sector collaboration that is needed to ensure the nation's ability to make progress toward achieving the vision articulated by the USCOP and the Ocean Action Plan (OAP).

Implementation Characteristics

To successfully address both near- and long-term ocean research priorities, several implementation aspects must be considered.

Use of Existing Mechanisms

Many of the mechanisms required to address effectively the ocean research priorities currently exist (see Mechanisms for Collaboration). Capitalizing on appropriate existing mechanisms will streamline implementation by reducing redundant organizing efforts, helping develop and maintain collaboration among diverse parties, and helping ensure that the majority of available resources are dedicated to addressing the research priorities and realizing research results.

Partnerships

Partnerships among local, state, tribal, regional, federal, and international bodies, as well as academic institutions, nongovernmental organizations, and industry, are required to successfully develop, enhance, and expand current and future research projects in ocean science and technology. These partnerships will enable leveraging of limited resources across sectors, build user capacity, and promote connectivity among diverse groups and regions.

Merit-Based Peer Review

Independent peer review promotes rigorous, objective, and equal treatment of all proposals for addressing the ocean research priorities. Although this process may not apply with the same weight to all scientific pursuits, such as those in mission-oriented or operational agencies, the merit-based peer-review process enables investments in projects that couple the best ideas and the most capable members of the community.

Sustained Efforts versus New Development

New initiatives and assets may be required to address gaps in knowledge or to take advantage of significant technological advances. However, many existing programs and infrastructure also address the research priorities. An appropriate balance of sustained existing efforts and new developments must be maintained for research efforts and the infrastructure that supports them.

National Priorities, Scaled Implementation

The development of ocean research priorities has focused a national effort on pursuing critical areas of ocean research necessary for the health and welfare of the nation. Successful efforts to address these priorities must account for differences in scales of the research efforts and research needs.

National implementation

Addressing certain research priorities requires national and/or federal effort. Many mechanisms to pursue this type of priority currently exist in the federal system and can be enhanced through the expanded agency collaboration under the Committee on Ocean Policy (COP) governance (see Mechanisms for Collaboration).

Regional, state, Tribal, and Local implementation

The national priorities outlined in this document need to be addressed at global to local scales and tailored to address differences among geographic regions, as well as different ocean uses, interactions, and phenomena within these areas.

Roles

The success of the effort to address the priorities described in this document requires the effective participation of not only federal agencies, but of numerous organizations and entities that exist in the United States and around the world.

Federal Agencies

Recognizing that no one group can fully address the ocean research priorities on its own, the role of the federal agencies is to provide the foundation of coordination, support, and resources to make progress on the priorities, consistent with the respective agency missions, and provide the impetus for collaboration among all sectors.

Local, Tribal, State, Regional Governance

Participation of governance entities at local, tribal, state, and regional levels will help ensure that efforts to address the national ocean research priorities are tailored to the needs of specific areas or groups, as appropriate. This focus can include the establishment of partnerships to address watershed or coastal issues affecting multiple states, such as accurate storm-surge and coastal-inundation forecasts. These governance entities also enable the effective use of results from nationally coordinated research efforts, such as hazards observations (e.g., hurricanes, tsunamis), by state and local agencies to advise the public and help mitigate impacts.

International Entities

International entities coordinate and implement global ocean research, management, and policy among the United States, other nations, and international groups. International entities also play a role in collaborating with the United States on all matters of regional oceanography (e.g., issues that cross national borders) and can influence research on natural hazards, ecological and human health, and fundamental processes affecting the Earth system.

Research Institutions

Research institutions provide a venue for scientific initiatives and collaborations among disciplines and sectors. These institutions can help enable close collaboration between researchers and stakeholders to better incorporate stakeholder issues into the research planning process, identify and coordinate synergistic research opportunities (e.g., shared data) among ongoing efforts, and identify and disseminate research findings to appropriate user groups.

Educational Institutions

Educational institutions communicate research results to the public and provide a foundation for informed discussion about society's interaction with the ocean. Educational institutions, ranging from the public and private K–16 educational providers, to universities, to informal venues such as aquaria, also provide a foundation for educating future researchers, managers, decision-makers, and leaders to expand understanding of the marine environment and to ensure the safe, effective, and sustainable future use of ocean, coastal, and Great Lakes resources.

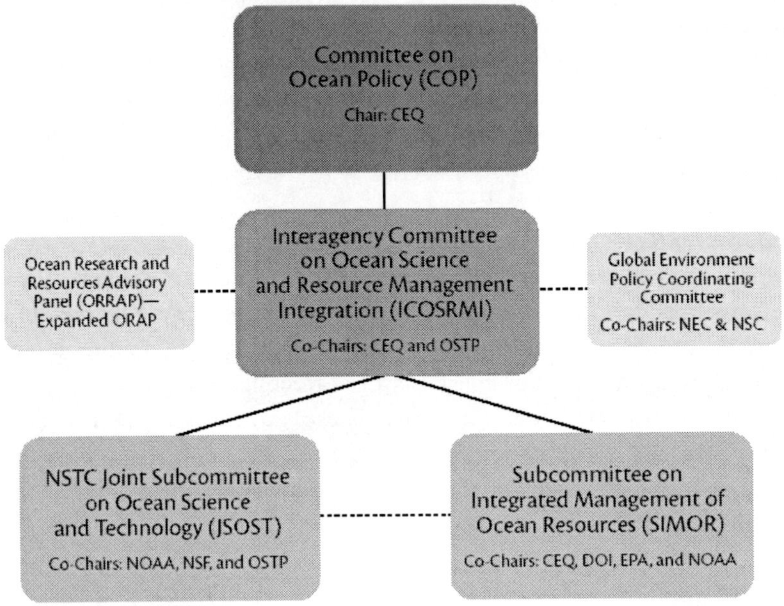

Figure 2. Cop Governance Structure.

Private Sector

The private sector provides expertise, resources, products, and services; uses research results in establishing economic development opportunities; and facilitates the development of research infrastructure. Developing research strategies in conjunction with the private sector will help ensure that these strategies leverage private-sector expertise, capability, and resources.

Nongovernmental Organizations

Nongovernmental organizations, such as advocacy groups and philanthropic organizations, are critical links among communities, whether representing and communicating needs, distributing resources, acting as conduits for research to enhance ocean literacy, or serving as stewards for the ocean ecosystem. Engagement of these organizations in developing plans to address high-priority ocean research is essential to ensure that diverse sectors have the opportunity to communicate and engage with each other.

Mechanisms for Collaboration

Among Federal Agencies

Collaboration and coordination among federal agencies was expanded with the establishment of the new governance structure under the COP (Figure 2). This new structure, comprised of representatives from the 25 federal agencies with interests or activities in the ocean, includes several subgroups that allow the federal agencies to coordinate efforts on issues such as science and technology (JSOST), resource management (Subcommittee on Integrated Management of Ocean Resources [SIMOR]), and international ocean issues

(Ocean Sub-Policy Coordinating Committee [Oceans Sub-PCC]). Efforts within these groups are further defined along topical areas, including ocean education, partnerships, and observations. Other ocean-related interagency collaborations focus on topical or geographic areas (e.g., Committee on the Marine Transportation System, U.S. Coral Reef Task Force, Great Lakes Interagency Task Force) and can be accessed to facilitate efforts to address the ocean research priorities. The JSOST and SIMOR, under the integrating auspices of the Interagency Committee on Ocean Science and Resource Management Integration (ICOSRMI), are leading the effort to coordinate federal-agency activities. These activities will include coordinating with other federal initiatives (e.g., U.S. CCSP, U.S. Group on Earth Observations), expanding interaction with other sectors, promoting collaborations among sectors, and advancing connections with regional initiatives.

Within Regions

Mechanisms for collaboration within geographic regions exist on the federal scale as well as on the regional, state, and local scale. These mechanisms include forums for interaction between federal agency representatives and state officials and resource managers (e.g., SIMOR Federal-State Science and Management Integration Task Team [FSSMITT], Coastal States Organization) on issues that have regional, state, or local relevance, such as the ocean research priorities.

Key mechanisms for collaboration will be derived from existing regional, state, tribal, and local entities, and where needed, new mechanisms may be established to address regional ocean and coastal issues. It is critical that collaboration mechanisms at all levels of government remain flexible enough to address specific needs at the appropriate scale. For example, regional consortia have been used to define a set of regional challenges, set regional research priorities based on regional needs and national research priorities, and develop and enact regional research implementation plans. Although no single regional structure can be applied to all cases, several existing or emerging organizations demonstrate key aspects to consider. The Sea Grant program encourages regional collaborations among diverse stakeholders through regional science planning efforts. The Chesapeake Bay Program is a watershed-based initiative designed to address ecosystem goals for the Chesapeake Bay and is supported by collaborations among local, state, and federal entities. Organizations oriented around ecosystem-based areas (e.g., Large Marine Ecosystems), such as Regional Fishery Management Councils (RFMCs) and Integrated Ocean Observing System (IOOS) Regional Associations, allow federal and regional interests to be represented and can help further ecosystem-based management. Still others are based on the common needs of groups of states, such as the Gulf of Mexico Alliance, the Northeast Regional Ocean Council, the Great Lakes Regional Collaboration, and the recent West Coast Governor's Agreement on Ocean Health, which represent local and state interests, in current or future collaboration with federal partners.

By promoting regional efforts through the coordination of existing regional efforts or the development of new regional consortia, priorities critical to the stakeholders in those regions can be identified and addressed. To ensure productive regional-federal collaboration, the JSOST, SIMOR, and ICOSRMI will work to ensure effective collaboration between the federal agencies and these regional consortia. Research efforts will be coordinated through a liaison system in which federal agency personnel in each region will interface with regional groups.

With Other Sectors

Several mechanisms currently exist that facilitate interaction and partnerships among multiple sectors. These mechanisms include: formalized interagency activities designed to identify and fund efforts that demonstrate public-private partnerships in specified areas of interest (e.g., National Oceanographic Partnership Program proposal solicitation program, now under the auspices of the JSOST Interagency Working Group on Ocean Partnerships [IWG-OP]) or to pool capabilities, expertise, and information among sectors to address local problems (e.g., Coastal America); stakeholder focus groups and roundtables that bring together parties with vested interest in specific topics or activities (e.g., SIMOR stakeholder focus groups, industry roundtables); and governmental advisory groups (e.g., Ocean Research and Resource Advisory Panel [ORRAP]). Although many of these efforts are initiated at the federal level, they represent forums for discussion and interaction between different groups. The JSOST will actively support and use existing activities under its purview that promote collaborations across sectors and will coordinate with the SIMOR and ICOSRMI to integrate existing federal mechanisms into the COP structure.

With International Groups

By capitalizing on and expanding international efforts, such as international treaties, bilateral agreements, and international scientific forums, the nation can more effectively pursue global research efforts that extend outside its borders, both in terms of influence and impact. Expanding federal interaction with international groups, such as the Intergovernmental Oceanographic Commission, the World Meteorological Organization, the International Maritime Organization, the International Hydrographic Organization, and the Intergovernmental Panel on Climate Change, will also promote effective two-way communication. The JSOST will coordinate with the Oceans Sub-PCC to further engage international ocean governance groups to promote international research efforts in support of the ocean research priorities.

With Resource Managers

Coordination between resource managers and ocean scientists will help ensure that the research needs for effective resource management are met and that the research results are applied by resource managers[37]. The ICOSRMI provides a venue to integrate ocean science and management at the federal level. The SIMOR, sister committee to the JSOST, is currently leading the federal effort to develop mechanisms for such scientist-resource manager coordination. One of these mechanisms is the establishment of task teams consisting of resource managers and researchers dedicated to improving two-way communication between these two communities (e.g., FSSMITT). Communication between resource managers and scientists can also occur as part of nationally sponsored regional outreach efforts (e.g., Sea Grant) and regional councils or groups (e.g., IOOS Regional Associations). Topical or region-oriented activities (e.g., National Dredging Team, coastal-zone conferences) or groups (e.g., RFMCs) can also be used to place new research results into the hands of resource managers, while providing a forum for resource managers to express their research needs and coordinate closely with the research community.

New Mechanisms for Collaboration

Although existing mechanisms for coordination should be used wherever possible, new mechanisms may be required to enable effective partnerships and collaboration. For areas where regional associations are not currently in place, the need for, and development of, a regional mechanism will be examined. Currently, there is no consistent and systematic set of coordinating mechanisms to develop broad-based regional research structures that would provide effective, ongoing regional implementation. The JSOST, working with the SIMOR, will guide a process leading to such a structure.

Additional new mechanisms for other sectors may include industry-university cooperative research centers that promote long-term partnerships among industry, academic institutions, and government, among others. The JSOST IWG-OP, working with its COP partners (e.g., other interagency working groups, JSOST, SIMOR), will identify areas that require new mechanisms for collaboration (e.g., technology development) and develop mechanisms appropriate for those sectors.

Infrastructure Needs

The priorities identified within each societal theme include some discussion of the tools critical to the success of the ocean research priorities. Additionally, elements common to many of the research priorities (e.g., ocean observing and models) are highlighted, along with a series of focused efforts necessary to achieve the full capabilities of these common elements. Central to effectively developing these necessary tools is an assessment of the many infrastructure and technology issues associated with the national ocean research priorities, such as a determination of current federal and national assets, recognizing that no one sector can fulfill all of the infrastructure needs. Development of this assessment will capitalize on the results of existing infrastructure assessments (e.g., fleet renewal plans) and projections of infrastructure enhancement or development (e.g., IOOS, Ocean Observing Initiative [OOI]).

Research Translation

Research translation—transforming research results into readily understandable and usable information, whether for decision-makers, resource managers, educators, or potential workforce participants—requires a common set of skills, currently embodied in many existing organizations. Building on existing mechanisms will be a critical component of moving research results into the broader community for use in education, management, and decision-making processes. Some organizations have begun to work together on topics such as ocean observing and workforce development, and provide a foundation for prototyping broader activities for research translation and expanding such activities.

Decision Support

Translating research results into decision-support and information products will arm resource managers and other decision-makers with the most accurate, research-based information possible. Mechanisms to do this translation already exist in some cases, and can be adapted and expanded. Federal entities, such as the National Oceanic and Atmospheric Administration (NOAA) Coastal Services Center, currently assist the nation's coastal-resource management and emergency-management communities by providing access to information, technology, and training, and by producing new tools and approaches that often are applied nationwide. State and local organizations, as well as regional councils and associations, can also serve to provide data, information, and products on marine and estuarine systems deemed necessary by user groups.

Future efforts related to decision support should consider a number of elements, many of which can be modified as needed and applied to education efforts. A good example of elements that should be considered for decision support is outlined in the Strategic Plan for the U.S. CCSP[38]. These elements are:

- Analyses structured around specific questions with consistent communications among all involved scientists and stakeholders addressing the questions
- Early and continuing involvement of, and feedback from, stakeholders, including resource managers, realizing that time and funding to attend workshops is severely limited
- Scientific syntheses and assessments that support informed discussion of ocean issues by decision-makers, stakeholders, the media, and the general public
- Adaptive management for resources and infrastructure
- Information resources that support adaptive management and planning and transfer these resources from research to application
- Methods (scenario evaluations, integrated analyses, alternative analytical approaches) that support policy-making

Research to Operations

Translating research capabilities to operational status is a critical concern for the ocean community, but its associated issues apply to science and industry across the board. Close collaboration with private-sector partners and identification of operational requirements is key to the success of transitioning research. Additionally, successful transitions will build on the tools and mechanisms of partnering (including with other research sectors outside of the ocean community), translation, and workforce development, outlined above.

Ocean Education

A variety of efforts exist to promote and support ocean education. Some entities connect ocean scientists with educators (e.g., Centers for Ocean Sciences Education Excellence). Others excel at introducing ocean-related concepts to the public (e.g., National Estuarine Research Reserve System), while still others focus on community-based outreach. The academic community is responsible for preparing a workforce with the interdisciplin-ary capabilities and intercommunity knowledge required to address the ocean research priorities. Coordination of education efforts at the local, state, and federal levels is key to establishing an ocean-literate nation.

Understanding how ocean science research and resource management communities can best contribute to improving formal education necessitates working in concert with the Academic Competitiveness Council, a federal group tasked to develop an inventory of federally funded science, technology, engineering, and mathematics (STEM) education programs and evaluate the effectiveness of these programs. The Interagency Working Group on Ocean Education, an entity under the auspices of the JSOST and SIMOR, provides the ideal forum in which to enhance ocean education in conjunction with broader national STEM efforts. These activities will also provide the foundation for a coordinated effort to develop and promote a comprehensive education message about the ocean and its role in the Earth system, and to enable the use of ocean-observing data for management and educational purposes. Such an effort would also help ensure that education-related translation efforts are more effective, that existing ocean science education networks and their ties to regional collaborations are strengthened, and that these individual activities are coordinated.

Developing the intellectual capital to pursue the research priorities requires an assessment of the current technological and scientific (natural and social sciences) workforce to understand gaps, surpluses, and critical skills. This baseline assessment will help identify areas that need to be developed or enhanced to provide the necessary workforce, and provide guidance to universities and other educational institutions about how best to prepare students for current and future workforce demands. Efforts to develop and expand the ocean science workforce include expanding opportunities for obtaining graduate degrees and interdisciplinary training, including experience-based ("hands-on") learning opportunities, developing professional certification programs, and creating competitive incentives to develop and establish training programs.

Assessment and Evaluation

Achieving the goals for each ocean research priority and ensuring that technology and infrastructure needs are met requires coordination among many research efforts. Thus, any regular assessment and evaluation of progress should have a performance-measurement structure that is comprehensive and reflective of multi-agency, multidisciplinary research efforts. The JSOST will consider both a large-scale evaluation of the effort as a whole as well as assessments of individual priority efforts.

Given the multi-agency scope of the ocean research priorities at the federal level, the JSOST intends to establish an umbrella mechanism for federal performance assessment of efforts to address the research priorities. Part of this mechanism will include evaluating the effectiveness of existing mechanisms for interagency and agency-sector (e.g., academia, resource management, nongovernmental organization, private sector) collaborative efforts. The performance-assessment structure in place for the multi-agency, multi-year U.S. CCSP is a useful reference for the development of this umbrella mechanism. The performance assessment will be reported to the ICOSRMI on an annual basis. Results of this assessment will also be used to guide the development of annual implementation activities. The external review that will occur as part of the five-year plan update (see Budget and Plan Update Mechanisms section) will also include an assessment of the effectiveness of the performance-assessment mechanism and an assessment of the success of the implementation.

The foundation of programmatic performance accountability at the federal level for the Ocean Research Priorities Plan and Implementation Strategy will originate from the federal Government Performance and Results Act, which lays out clear guidelines for measuring program success. Milestones and metrics (qualitative and quantitative) that identify targets to be achieved and assess and determine outputs, outcomes, and impacts prior to specific implementation activities will be the centerpiece of an assessment and evaluation system. Performance goals for federal agencies should be consistent with the overarching goals of the individual priority as well as individual agency missions.

Budget and Plan Update Mechanisms

Many aspects of this implementation strategy will be useful for other sectors as they seek to implement the ocean research priorities. This section focuses specifically on federal budgetary aspects and federal actions to update the ocean research priorities.

Budget

Federal efforts to address the ocean research priorities should be considered during the development of the President's budget, as these interagency efforts will greatly increase the effectiveness of federal activities when carried out in a coordinated and integrated fashion.

In consultation with the SIMOR and its partners and upon approval by the ICOSRMI, the JSOST shall produce a memorandum articulating the interagency priorities for the fiscal year (FY) under consideration to inform the annual federal budget process. The memorandum shall reference the specific interagency approaches to address the priorities, the nature of the programs to address each priority, and the partners to participate in efforts to address the priority. Cost estimates and funding options for the federal component of each effort will be determined separately for internal consideration. The memorandum shall be developed and forwarded to relevant federal departments and agencies approximately 22 months before the beginning of the fiscal year (e.g., the guidance memorandum for FY2010 will be released in January of 2008). The JSOST will work to incorporate their key recommendations in the joint Office of Science and Technology Policy (OSTP)-Office of Management and Budget (OMB) annual research and development memorandum. As the JSOST works to incorporate their key recommendations into the OSTP/OMB annual memorandum, the JSOST will also develop and implement a collaborative outreach effort to Congress, states, and stakeholders to advance the national ocean research priorities.

Plan Update

A successful effort to coordinate research nationwide must have adequate flexibility to address emerging challenges and to take advantage of opportunities provided by scientific and technological breakthroughs, while providing the context that comes from development of a long-term perspective.

The federal agencies, through the JSOST, will annually review the national ocean research priorities as part of the ongoing interagency planning and coordination activities undertaken by the COP groups. This annual review of the ocean research priorities will include the near-term implementation steps for the COP groups to address the priorities (both near-term and long-term). Near-term priorities (2–5 years) will be refreshed approximately every five years. On an intermediate frequency (~ 3 years), the JSOST (in cooperation with the SIMOR) will hold "consultation" workshops with major non-federal sponsors, participants, and end users of ocean science and technology development to identify opportunities to enhance the impact of federal research through strategic partnerships. The JSOST will conduct a formal review of the broad, national ocean research priorities every five years in a manner consistent with the development of the current report (e.g., internal assessments, public workshops, public-comment periods, outreach efforts). Such a frequency would allow long-term (decadal) priorities to be revisited and possibly revised. This update will also include a review of the implementation mechanism on the part of the federal government and an assessment of the success of the implementation of the priorities. An external review of the updated priorities and implementation mechanisms for the federal agencies and federal partners will also occur at this time.

Addressing Near-Term Priorities

Initial plans have been developed to address the current set of near-term priorities. Although these plans were developed by the federal agencies, successful implementation relies on the coordinated efforts of a variety of sectors, including state agencies, regional initiatives, academic collaborations, and industry partnerships.

Forecasting the Response of Coastal Ecosystems to Persistent Forcing and Extreme Events

The objective of this near-term priority is to provide managers and decision-makers with the tools needed to prepare for and respond to extreme weather events, natural disasters, and changing natural and human influences. This activity will require the development and integration of predictions of environmental conditions with a full assessment of the vulnerability and resilience of coastal ecosystems and communities.

Approach

This effort will focus on several regional systems and support end-to-end development and integration of observations, research, and forecast models leading to decision-support tools. Research on ecosystem and community conditions, processes, and responses to hazards and forcings, supported by observations, will enhance forecast capabilities and provide the necessary inputs for model development. Enhancement and integration of existing observing capabilities and infrastructure will ensure system interoperability and availability of critical observational data required to support model development, operational forecasts, and decision-support tools. These activities will leverage and advance national efforts to develop and integrate observational and modeling systems while addressing regional needs. This effort will coordinate and integrate existing observational, research, and monitoring programs from several federal agencies, reflecting their mission-focused roles and existing programs. Agency contributions necessary to address this priority include:

- ***United States Geological Survey***—Conduct hydrological and biological monitoring and regional geologic and environmental mapping and characterization. Perform research on coastal processes and response to persistent forcing and extreme events. Develop models to forecast hydrologic, landscape, and ecosystem response, and tools to provide assessments of coastal vulnerability and predict future ecosystem conditions.

- *NOAA*—Acquire and integrate (including standards development) monitoring and mapping data from existing and enhanced coastal observation platforms. Conduct assessments to identify region-specific coastal and marine research priorities. Develop community inundation and ecosystem models for assessing storm vulnerability, oil-spill movements, and ecological impacts. Develop a Web-based geospatial framework and digital-elevation models essential for decision-support tools, including socioeconomic and environmental indices.
- ***United States Army Corps of Engineers***—Expand shallow-water coastal mapping and wave-observation programs. Develop community, high-resolution coastal models and testbeds to support next-generation models for all agencies, and transition models from research to operational use. Enhance coupling among ecosystem, coastal hydrodynamic, and watershed models to provide integrated management tools for planning, hazard identification, and response.
- ***Environmental Protection Agency (EPA)***—Integrate and assimilate observations from coastal-condition surveys and observational platforms. Enhance coastal-condition surveys and assessments. Develop ecosystem models to forecast changes in ecosystem services. Develop models and decision-support tools, including socioeconomic measures of goods and services, to assess changes in ecosystem services resulting from regulatory programs, land-use planning, and natural events.
- ***National Science Foundation (NSF)***—Generate real-time, reconfigurable, open-ocean observations that may be used to properly initialize forecast models through the OOI. Conduct research on physical and ecological processes relevant to model development and basic understanding of ecosystem response.

Collaborative Efforts

Region-specific priorities and implementation strategies to address those priorities will result from engagement across a variety of sectors. Federal, state, and academic research and user communities will join in planning and implementation; agencies will cooperatively execute the resulting suite of priority projects.

Comparative Analysis of Marine Ecosystem Organization

The objective of this near-term priority is to improve the management and sustainability of marine ecosystems by better quantifying ecosystem dynamics in relation to human activities. This improved quantification will be used in evaluating the effectiveness of MPAs in achieving management objectives. MPAs incorporate an array of managed marine ecosystems with differing levels of protection, from areas that allow multiple uses to those that restrict take and/or access. A better scientific understanding of the function and impacts of MPAs will assist in making decisions regarding alteration of current MPAs and potential designation of new U.S. MPAs.

Approach

Three primary research activities will address this near-term priority. First, new modeling and analysis procedures will be developed to provide the foundation for comparative analyses and evaluation. This development will also address a key challenge in providing reliable forecasts of biological phenomena supporting ecosystem-based management: the paucity of information on mechanisms controlling complex behaviors within ecosystems. Second, using these quantitative methods, data collected by agencies and academic scientists will be consistently applied to selected ecosystem types in order to provide improved understanding of ecosystem response to environmental change. Third, fieldwork, analysis, and modeling are needed inside and in adjacent areas outside selected MPAs to understand their impacts and evaluate their success on several scales. Agency contributions necessary to address this priority include:

- *NSF*—Improve ways of incorporating uncertainty in model-parameter estimates, the role of size and age structure in population demographics, and spatial dynamics. Facilitate application of quantitative frameworks to data sets to synthesize dynamics across ecosystems and conduct investigations with theory, design, observations, and experiments to interpret the ecosystem and socioeconomic impacts of MPAs.
- *NOAA*—Develop quantitative ecosystem models in partnership with academic and agency partners, apply appropriate data from multiple in-house monitoring databases, and interpret results. Analyze the MPA role in relation to conventional fisheries management tools, in addition to other ecosystem variables.
- ***National Aeronautics and Space Administration (NASA)***—Expand existing biological and biogeochemical models to encompass the function of ecosystems under study, and facilitate incorporation of satellite ocean biological and biogeochemical data into models to improve spatial ecosystem-function analyses. Examine quantified variations in key ocean properties from satellite data that may influence the amount and composition of primary productivity in and around MPAs.
- ***Department of Interior (DOI)***—Contribute quantitative approaches for species assessments and ecosystem evaluation, coral reef, deep benthic and wetland/estuarine ecosystem data for the models, and relevant seafloor mapping data and analyses, especially where specific human activities that impact ecosystems are affected by the imposition of MPAs.

Collaborative Efforts

The federal agencies and their partners will select a set of geographical regions and MPAs that represent an optimal mix of system types to study. Physical and biological data sets collected by participating scientists and agencies will be used in models developed by the program. This effort will build upon existing academic programs such as Global Ocean Ecosystem Dynamics and the Long-Term Ecological Research network. Partnerships with these and other groups, such as the U.S. Coral Reef Task Force, RFMCs, and state and federal coastal management agencies, will be used to achieve priority objectives and ensure that the research results are used in current management and future area designations.

Sensors for Marine Ecosystems

The objective of this near-term research priority is to improve understanding of ocean ecology on multiple spatial (from individual cells to the global ocean) and temporal (from seconds to decades) scales by improving sensor capabilities and their associated time- series measurements of ocean biological (e.g., genomic), biogeochemical, chemical, and bio-optical properties. Synthesizing multi-scale oceanographic observations with existing data will provide a new way of "seeing" and analyzing ocean-ecosystem functions. This new view will allow researchers to better understand ecosystem response to environmental and climate variability and change, resulting in improved ecosystem-management strategies and protection of public health.

Approach

Three major ecological spatial scales (from cells to ocean regions to the global ocean) need to be observed in a single snapshot, and the information integrated over multiple time scales, to attain the next level of understanding of ocean-ecosystem function. Immediate improvements can be made to *in situ* instrumentation in eco-genomics and interdisciplinary ocean observing, and in optical and bio-optical sensors in support of existing satellite ocean-color data and new space-based biological and biogeochemical observations. Such advancements will provide information at temporal and spatial scales not presently available through any one agency's observing capabilities. Agency contributions to address this priority include:

- *NSF*—Develop *in situ* ocean sensing systems, emphasizing chemical sensors for high-priority compounds such as nutrients, biological sensors for real-time ecosystem analysis, and physical sensors that enable better understanding of transport and fluid velocity. Develop associated technologies such as cyberinfrastructure tools (to link sensors in a broad network context) and advanced power-generation capabilities.
- *NASA*—Improve existing space-based, open-ocean, remote-sensing capabilities of biological and biogeochemical properties (and future satellite sensor capabilities in optically complex or coastal waters) by assessing accuracies in remote-sensing, optical, and bio-optical sensor calibration and validation data-collection activities. Develop next-generation optical and bio-optical field sensors and satellite data products that test new technological and methodological approaches.
- *NOAA*—Develop genomic libraries to advance understanding of ecosystem processes, as well as species abundance and distribution. Develop *in situ* sensors for rapid detection of pathogens, harmful algae, and toxins, and methods to integrate biosensor data with other ocean observations. Develop genomic and proteomic tools and supporting bioinformatics infrastructure to elucidate effects of multiple environmental stressors on marine organisms. Improve video plankton recorders for recruitment process studies.

Collaborative Efforts

Improving existing *in situ* sensors and developing new sensors requires the participation of engineering and research communities. Research communities will validate the new and improved measurements. Application of the data gained from these advances in sensors will be undertaken through the coordinated efforts of researchers and resource managers. Genomic, *in situ* biological, chemical, and bio-optical field observations, remote sensing, and numerical models will be combined into diagnostic tools for regional to global investigations and linked with the developing coastal and offshore observatory efforts to maximize stakeholder participation, input, and use of the advances. Communities will be engaged through a variety of mechanisms, including planning workshops and peer-reviewed research solicitations.

Assessing Meridional Overturning Circulation Variability: Implications for Rapid Climate Change

The objective of this near-term priority is improved understanding of the mechanisms behind fluctuations of the MOC, which will lead to new capabilities for monitoring and making predictions of MOC changes. These predictive capabilities will help identify the impacts of MOC changes on the ocean, climate, extreme weather events, regional sea- level changes, ecosystems, and carbon budgets.

Approach

Given current limited understanding of the MOC, describing the MOC, its variability, and critical processes, and improving the ability to model these will be an early and ongoing emphasis of the program. Enhanced understanding of the MOC system, including its processes and fluctuations, will be gained through the analysis of existing data sets. This expanded understanding will provide the foundation for the design and implementation of a comprehensive MOC observation and monitoring program. In parallel, new forecasting capabilities will be developed by improving ocean models, coupled models, and ocean analyses for their initialization. Finally, interdisciplinary research and modeling studies will be required to characterize the impacts and feedbacks of changes in the MOC on ecosystems, carbon budgets, and regional climate. Initial federal implementation is derived from the work of three agencies and has many linkages with CCSP. Agency contributions to address this priority include:

- *NASA*—Use current and future ocean measurements from satellites to assess processes responsible for MOC variability and to improve ocean models through advanced ocean-state estimation. Refine ocean-state estimates spanning the past 50 years, incorporating all available *in situ* and remotely sensed observations. Use modeling studies to assess impacts of MOC changes on sea-level changes and high latitudes.
- *NOAA*—Conduct modeling experiments on the origins of MOC variability leading to improved understanding of the relative roles of wind and thermohaline forcing. Test models and theories against climate data sets and implement improved models for MOC-related studies.
- *NSF*—Conduct relevant ocean-process studies in the Atlantic and sub-Arctic Oceans to improve ocean-model parameterizations, and perform MOC analysis and modeling studies. Support historical MOC reconstructions and improved data- assimilation systems; initiate modeling studies of the impacts of MOC changes on North Atlantic storminess, ecosystems, and ocean carbon uptake.

Collaborative Efforts

This activity is founded on the coordinated efforts of university, federal, and international initiatives to understand and forecast this large-scale ocean phenomena and its potential for global impacts. This activity will leverage and support the Administration's CCSP priority in "*Assessing abrupt changes in a warming world.*" It builds on ongoing collaboration with international partners such as the United Kingdom's Rapid Climate Change program and the European Union Developing Arctic Modeling and Observing Capabilities for Long-term Environmental Studies program, as well as developing South American partnerships targeting observing needs for the southern hemisphere branch of the MOC. It takes advantage of national and international ocean remote-sensing satellites, which are part of GEOSS. Activities will be established to engage end-user communities to assess the impacts of decadal climate variability on their decision-making processes and to identify future product suites that would provide needed information.

ACRONYMS

CCSP	Climate change science Program
CEQ	Council on Environmental Quality
CO_2	Carbon dioxide
COP	Committee on Ocean Policy
DOI	Department of interior
EEZ	Exclusive Economic Zone
EPA	Environmental Protection Agency
FSSMITT	Federal-state science and Management integration Task Team
FY	Fiscal year
GEOSS	Global Earth Observing system of systems
GIS	Geographic information system
HAB	Harmful algal bloom
ICOSRMI	Interzagency committee on Ocean science and Resource Management integration
IOOS	Integrated Ocean Observing system
IWG-OP	Interagency Working Group on Ocean Partnerships
JSOST	Joint subcommittee on Ocean science and Technology
MOC	Meridional Overturning circulation
MPA	Marine Protected Area
NAO	North Atlantic Oscillation
NASA	National Aeronautics and space Administration
NEC	National Economic council
NOAA	National Oceanic and Atmospheric Administration
NOPP	National Oceanographic Partnership Program
NSC	National security council
NSF	National science Foundation
OAP	Ocean Action Plan
OCEANS SUB-PCC	Ocean sub-Policy coordinating committee
OCS	Outer continental shelf
OMB	Office of Management and Budget
OOI	Ocean Observatories initiative

ORRAP	Ocean Research and Resource Advisory Panel
OSTP	Office of science and Technology Policy
PDO	Pacific Decadal Oscillation
RFMC	Regional Fishery Management council
SIMOR	Subcommittee on integrated Management of Ocean Resources
STEM	Science, technology, engineering and mathematics
USCOP	U.S. Commission on Ocean Policy

ACKNOWLEDGMENTS

This report has been developed over two years, through the dedicated effort of many valuable contributors. We are grateful to all interagency committee on Ocean science and Resource Management integration (icOsRMi) and committee on Ocean Policy (cOP) members for their strong support and constructive comments throughout the process of developing this report. We also recognize and sincerely appreciate the helpful efforts of the subcommittee on integrated Management of Ocean Resources (siMOR), its Federal-state Task Team (FsTT) and the Ocean Research and Resources Advisory Panel (ORRAP).

Hundreds of individuals have written, edited and commented upon this report. We particularly acknowledge all those who participated in the public workshop and the community town hall meetings and those who responded during the public comment periods. special thanks go to those who participated in the National Research council (NRc) review of this document. To all of these contributors we express our appreciation. There are several individuals, in particular, who have delivered specific time and effort to this process. They deserve special citation and gratitude: shelby Walker, for tireless attention to writing and editing; Lynn Dancy, Kristen Koch, Emily Larkin, Frank Lockhart, sheridan MacAuley, Roxanne Nikolaus, Kenli schaaf, Laura snow, and Anne Tenney for outstanding technical support; and the JsOsT members during the development of this report (see following list).

End Notes

[1] The term "ocean," unless explicitly noted in the text, refers to the open ocean, coasts (including bays and estuaries), coastal watersheds, and Great Lakes.

[2] "Ecosystem-based management is an integrated approach to management that considers the entire ecosystem, including humans." Scientific Consensus Statement on Marine Ecosystem-Based Management. March 21, 2005. http://compassonline.org/?=EBM (accessed January 24, 2007)

[3] U.S. Commission on Ocean Policy. 2004. *An Ocean Blueprint for the 21st Century, Final Report.* Washington, DC, p. 378

[4] For example, National Research Council. 2003. *Exploration of the Seas: Voyage into the Unknown.* National Academy Press, Washington, DC.; Brewer, P. and T. Moore, eds. 2001. *Ocean Sciences at the New Millennium.* UCAR/JOSS.

[5] Approximately 3.4 million square miles (http://www.uscg.mil/hq/g-o/g-opl/LMR/LMR.htm) (accessed January 24, 2007)

[6] U.S. Commission on Ocean Policy. 2004. *An Ocean Blueprint for the 21st Century, Final Report.* Washington, DC, p. 275.

[7] *Report to Congress: Comprehensive Inventory of the U.S. Outer Continental Shelf Oil and Natural Gas Resources.* Energy Policy Act of 2005, Section 357, U.S. Department of the Interior, Minerals Management Service, Table 2, page 31, February 2006.

[8] http://www.nrel.gov/docs/fy06osti/39479.pdf (accessed January 24, 2007)

[9] Collett, T.S. 1995. Gas hydrate resources of the United States. In: *National Assessment of United States Oil and Gas Resources*. D.L. Gautier, G.L. Dolton, K.I. Takahashi, and K.L. Varnes eds, available on CD-ROM: U.S. Geological Survey Digital Data Series 30.

[10] Collett, T.S. 1997. Resource potential of marine and permafrost associated gas hydrates, in *Oceanic Gas Hydrate: Guidance for Research and Programmatic Development at the Naval Research Laboratory.* M.D. Max, R.E. Pallenbarg, and B.B. Rath, eds, Proceedings of the workshop on Naval Research Laboratory gas hydrate research program, September 23–24, 1997, Washington, D.C. NRL/MR/6100-97- 8124, 51 pp.

[11] http://www.glerl.noaa.gov/pr/ourlakes/ (accessed January 24, 2007)

[12] U.S. Commission on Ocean Policy. 2004. *An Ocean Blueprint for the 21st Century, Final Report.* Washington, DC, p. 35.

[13] http://www.oceanservice.noaa.gov/programs/mb/supp_cstl_population l (accessed January 24, 2007)

[14] http://www.ncdc.noaa.gov/oa/reports/tech-report-200501z.pdf (accessed January 24, 2007)

[15] U.S. Commission on Ocean Policy. 2004. *An Ocean Blueprint for the 21st Century, Final Report.* Washington, DC, p. 192.

[16] Arctic Council and International Arctic Science Committee. 2004. *Impacts of a Warming Arctic: Arctic Climate Impact Assessment.* Cambridge University Press. Online available at http://www.acia.uaf. edu/pages/overview. html (accessed January 24, 2007).

[17] Short sea shipping uses inland and coastal waterways to move commercial freight from major domestic ports to its destination. (http://www.marad.dot.gov/Programs/sssbroc.htm) (accessed January 24, 2007)

[18] Research efforts for marine transportation will be coordinated with efforts of the cabinet-level Com- mittee on the Marine Transportation System (http://www.cmts.gov/; accessed January 24, 2007)

[19] http://www.uscg.mil/hq/g-o/g-opl/LMR/LMR.htm (accessed January 24, 2007)

[20] http://www.bts.gov/publications/pocket_guide_to_transportation (accessed January 24, 2007)

[21] http://www.natice.noaa.gov/icefree/FinalArcticReport.pdf (accessed January 24, 2007)

[22] For example, the Multipurpose Marine Cadastre Initiative (http://www.mms.gov/ld/PDFs/ MappingInitiative.pdf) (accessed January 24, 2007).

[23] U.S. Climate Change Science Program and Subcommittee on Global Change Research. 2001. *Strategic Plan for the U.S. Climate Change Science Program.* Washington, DC. Online available at http:// www.climatescience.gov/Library/stratplan2003/final/default.htm. (accessed January 24, 2007)

[24] Ecosystem health goals and objectives are management tools that guide an ecosystem-based approach to environmental planning and management. Ecosystem goals are statements that describe the desired state of an ecosystem. Ecosystem objectives are more specific. They describe desired conditions for a given ecosystem accounting ecological characteristics and uses.

[25] Millennium Ecosystem Assessment. http://www.maweb.org//en/About.Overview.aspx (accessed January 24, 2007)

[26] The study of the evolutionary relationships of species

[27] http://seafoodandhealth.org/documents/MicrosoftPowerPointPainter.pdf (accessed January 24, 2007); Scoging, A.C. 1992. Illness associated with seafood. *Canadian Medical Association Journal* 147(9):1,344–1,347.

[28] "Seafood" refers to all consumer seafood products as well as all forms of recreational and subsistence take, such as marine mammal harvest by Native Americans.

[29] Excluding allergic reactions to seafood

[30] Kuiken, T., F.A. Leighton, R.A.M. Fouchier, J.W. LeDuc, J.S. M. Peiris, A. Schudel, K. Stöhr, and A.D.M.E. Osterhaus. 2005. Pathogen surveillance in Animals. *Science* 309:1,680–1,681; Taylor, L.H., S.M. Latham, and M.E.J. Woolhouse. 2001. Risk factors for human disease emergence. *Philosophical Transactions of the Royal Society B:Biological Sciences* 356:983–989.

[31] U.S. Commission on Ocean Policy. 2004. *An Ocean Blueprint for the 21st Century, Final Report.* Washington, DC, pp. 340–342.

[32] Mozaffarian, D., and E.B. Rimm. 2006. Fish intake, contaminants and health. Evaluating the risks and benefits. *Journal of the American Medical Association* 296:1,885–1,899.

[33] http://www.ioc-goos.org/content/view/12/26/ (accessed January 24, 2007)

[34] http://www.aza.org/AboutAZA/CollectiveImpact1 (accessed January 24, 2007)

[35] Persistent forcing includes non-extreme, non-episodic influences, such as prevailing climatologic and oceanographic conditions. Persistent forcings include, for example, variations in nutrient and sediment inputs and ocean chemistry beyond those associated with extreme episodic events such as hurricanes.

[36] The number of managed ecosystems continues to increase. For example, in 2006, the United States has designated the world's largest marine conservation area in the Northwest Hawaiian Islands (http://www.whitehouse.gov/ news/releases/2006/06/print/20060615-18.html) and, more recently, the world's largest MPA in the Aleutian Islands (http://www.noaanews.noaa.gov/stories2006/s2673. htm). (accessed January 27, 2007)

[37] Subcommittee on Integrated Management of Ocean Resources. 2006. *Implementing the Work Priorities of the Subcommittee on Integrated Management of Ocean Resources (SIMOR): Work Plan.* p. 10 (http:// ocean.ceq.gov/about/docs/SIMOR_WorkPlan_Final.pdf) (accessed February 7, 2007).

[38] U.S. Climate Change Science Program and Subcommittee on Global Change Research. 2001. *Strategic Plan for the U.S. Climate Change Science Program*. Washington, DC. Online available at http:// www.climatescience.gov/Library/stratplan2003/final/default.htm. (accessed January 24, 2007)

In: Priorities for U.S. Ocean Science and Research
Editor: Ryan L. Murphy
ISBN: 978-1-61209-706-0

Chapter 2

INTERAGENCY OCEANS AND HUMAN HEALTH RESEARCH IMPLEMENTATION PLAN: A PRESCRIPTION FOR THE FUTURE

The Interagency Working Group on Harmful Algal Blooms, Hypoxia, and Human Health

EXECUTIVE SUMMARY

The ocean[1] provides numerous benefits to humankind, including seafood[2], pharmaceuticals and other natural products, recreational opportunities and aesthetic value, shoreline protection, waste assimilation, nutrient cycling, oxygen production, and drinking water[3]. The ocean also moderates climate and contributes substantially to the U.S. economy. However, the ocean poses risks when people are exposed to disease-causing organisms transmitted through or originating in the marine environment, toxins from harmful algal blooms (HABs) and other microorganisms, chemical contaminants, and when catastrophic events such as hurricanes and tsunamis occur.

Because the close connection between ocean ecosystem conditions and human health has long been recognized, several federal agencies have been working on oceans and human health (OHH)-related problems for many years. Ongoing activities and interagency collaborations have included programs related to seafood and drinking water safety, drug discovery, pollution effects and control, HABs and the toxins they produce, occurrence and transmission of disease-causing agents, and numerous others. However, the advent of named OHH programs, specifically the National Oceanic and Atmospheric Administration's (NOAA) Oceans and Human Health Initiative (OHHI) and the National Science Foundation (NSF) and National Institute of Environmental Health Sciences' (NIEHS) Research Centers of Excellence in OHH, has led to substantially increased progress in OHH research and its useful application, and in partnering among federal agencies, academia, states, and the private sector. Federal agencies collaborating on OHH research, in addition to NOAA, NSF, and NIEHS, include the Centers for Disease Control and Prevention (CDC), Environmental

Protection Agency (EPA), Food and Drug Administration (FDA), Marine Mammal Commission (MMC), National Aeronautics and Space Administration (NASA), and U.S. Geological Survey (USGS).

To coordinate ongoing OHH research efforts and to ensure that federal investments in this emerging and important interdisciplinary scientific field are utilized as efficiently as possible, in 2004 the U.S. Congress passed the Oceans and Human Health Act that formally established the Interagency OHH Program. The Act also explicitly calls for this ten-year implementation plan and states that its purpose is to define "the goals and priorities for Federal research which most effectively advance scientific understanding of the connections between the oceans and human health, provide usable information for the prediction of marine-related public health problems, and use the biological potential of the oceans for development of new treatments of human diseases and a greater understanding of human biology."

In fall 2005, the Joint Subcommittee on Ocean Science and Technology (JSOST) of the National Science and Technology Council (NSTC) Committee on Science and the Committee on Environment and Natural Resources and the Interagency Committee on Ocean Science and Resource Management Integration (ICOSRMI) established the Interagency Working Group on Harmful Algal Blooms, Hypoxia, and Human Health (IWG-4H). One of the IWG-4H's principal tasks has been to develop this implementation plan. This plan complements and expands upon the OHH-related priorities detailed in "*Charting the Course for Ocean Science in the United States for the Next Decade: An Ocean Research Priorities Plan and Implementation Strategy*" prepared by the JSOST and published in early 2007, and provides a framework for implementation of a national OHH research and outreach program. It also summarizes work underway in NOAA's OHHI and the NSF-NIEHS Centers in OHH, and provides synopses of ongoing related work in the CDC, EPA, FDA, MMC, NASA and USGS (see Chapter 2 for details).

In preparing this implementation plan, the IWG-4H focused on three broad goals of the Interagency OHH Program.

1) Establish a research agenda focused on the highest priority areas and activities, the appropriate mix of supporting infrastructure, and the transition of research to useful application and outreach.
2) Promote a robust interdisciplinary approach that links marine and biomedical sciences to create a fundamentally new OHH research community with the breadth of expertise required to address diverse human health questions within the complexity of ocean systems.
3) Establish and expand collaborative partnerships to develop and effectively use OHH information and products through technology transfer and training, and ensure inclusion of OHH-supportive data collection, integration, analysis, and modeling within the larger framework of national and international ocean observation efforts.

To accomplish our vision of a future where a healthy ocean allows people to benefit from a variety of ocean products and to swim, fish, or otherwise recreate in ocean waters and eat seafood without risk of exposure to disease-causing organisms and compounds, federal agencies and their academic and other partners should concentrate efforts where significant opportunities for advancement to improve and/or protect human health have been identified. These are:

1) Priority research dealing with pathogens, chemical contaminants, HABs, seafood safety, pharmaceuticals and other beneficial products and in cross-cutting areas such as epidemiology, sentinel species, genomics and related technologies, and social, behavioral and economic sciences.
2) Appropriate infrastructure to support research advances in such areas as linking to the ocean observing systems, data management and access, development of standards and standardized methods, and access to the sea.
3) Transition of research results to applications through targeted outreach and education activities and development of rapid response capabilities.

In addition, the IWG-4H identified the following six implementation actions that the federal agencies involved in the Interagency OHH Program will take to advance OHH research and application, leading to reduced health risks and increased health benefits for people.

Action 1. Work through Existing OHH Programs and Partnerships

Specifically work through existing NSF-NIEHS OHH Centers and NOAA's OHHI including OHHI Centers, external grants, scholars, and traineeships, and through collaboration with OHH-related activities in the CDC, EPA, FDA, MMC, NASA, and USGS.

Action 2. Target Priority Research in the Following Areas

- Basic and applied studies of ocean ecosystems and processes that affect human exposure to health risks, including natural hazards, seafood consumption, and climate change.
- Discovery and development of new pharmaceuticals and other beneficial products.
- Epidemiologic studies and disease surveillance to elucidate acute, chronic, and emerging health effects and threats in humans and animals resulting from exposures to ocean waters, sediments, air, and seafood.
- Use of marine species and habitats as early warning indicators of existing and emerging threats to human, animal, and ecosystem health and as models for the study of human disease processes and toxicology.
- Development and improvement of conceptual and quantitative models that integrate environmental, biological, and epidemiologic data to produce new levels of understanding and predictive capacity.
- Economic and socio-cultural studies that advance understanding of how humans use and value the health benefits provided by the ocean.

Action 3. Support Research Infrastructure Including

- Ocean observing, including development, deployment, operation and maintenance of biological sensors and other biological data-collection tools.
- Cyberinfrastructure to enable data sharing, integration, archiving, analysis, and access for a broad range of OHH-related data.
- New standards and standardized methods for OHH research and research materials.
- Access to remote and at-sea platforms for the OHH program to accomplish crucial sampling activities.
- Core facilities in genomics/proteomics, marine microbiology, and analytical chemistry.

Action 4. Support Transition of OHH Research to Application through Outreach and Education By

- Strengthening cross-agency, cross-institution and cross-discipline collaboration and coordination to provide the framework to develop a fundamentally new interdisciplinary OHH research community.
- Providing OHH information, tools, technology, products, and training to support improved public health and coastal and marine resource decision- making.
- Providing opportunities for interdisciplinary training, research, and collaborations for graduate and post-doctoral students, health professionals, and scientists at early- to mid-career stages.
- Coordinating OHH outreach and education programs across agencies.
- Supporting interagency partnerships to enable rapid and coordinated response of OHH researchers to emergency situations.
- Promoting ocean stewardship and ocean and human health literacy.

Action 5. Improve Coordination of OHH Activities within and across Agencies and Internationally

Much integration has been achieved by leaders and participants in the named NOAA and NSF-NIEHS OHH programs and related federal programs. However, for the Interagency OHH Program to continue to develop, long-term support for interagency coordination of OHH research and outreach is essential. To date, the IWG-4H has added appreciably to the positive interactions among OHH-related programs in a broad suite of agencies. The JSOST, through the IWG-4H, will continue to address interagency coordination issues.

Action 6. Provide for Updates to the Interagency OHH Research Implementation Plan

The OHH Act of 2004 requires that the Interagency OHH Annual Reports include a summary or copy of the implementation plan and any changes made in the plan. This provides a regular opportunity for the Interagency OHH community to update the plan.

These six actions will ensure a vibrant interagency program in OHH research and will result in better-informed policy decisions; discovery and application of natural products beneficial to humans; increased basic knowledge and enhanced understanding of human disease processes; improved understanding and communication about the benefits and risks of seafood consumption; increased economic returns due to decreased losses associated with beach and fishery closures, health care costs, and gains in the ocean natural products sector; and improved ocean stewardship and ocean literacy.

1: A Vision for Oceans and Human Health Activities in Service to Society

America's ocean[4] provides a strong foundation for the nation's economy and the well-being and health of its citizens. The ocean impacts every person in the United States no matter whether they are among the more than 50 percent of the population living along the coasts or those residing inland. The U.S. has over 95,000 miles of shoreline and 25,000 miles of navigable waters. Its Exclusive Economic Zone (EEZ) is the largest in the world with 3.4 million square miles of open water, and is larger than the landmass of the entire country. Flanked by the Atlantic, Pacific, and Arctic Oceans, the Bering and Caribbean Seas, and the Gulf of Mexico, and threaded with coastal watersheds, America's ocean serves as a highway for maritime transport, a supplier of seafood[5] and other natural products, and a prime tourist destination. America's coasts also provide essential services such as shoreline protection, nutrient cycling, waste assimilation, climate moderation, and a buffer against natural and man-made disasters. Overall, these waters help drive an economic engine that provides over half the nation's gross domestic product and more than 60 million jobs annually.

While the ocean provides invaluable benefits to people, effects on coastal and ocean ecosystems from multiple stressors including population growth, land use, loss of biodiversity, habitat degradation, pollution, and natural processes are leading to increased health threats to both humans and marine animals. Escalating climate change can exacerbate a number of these stressors, including the spread of disease, toxins, and contaminants in the marine environment, with polar areas perhaps especially vulnerable. Addressing these threats requires sustained and targeted efforts from multiple disciplines and institutions, development and deployment of new technologies, effective transfer of critical information to environmental and public health managers and the general public, and a focus on prevention and prediction. Such efforts can best be carried out through a coordinated interagency program as envisioned by the U.S. Commission on Ocean Policy(USCOP) and required by the Oceans and Human Health Act of 2004 (see Appendix 1).

Healthy beaches sustain a booming coastal tourism industry, attract over a 100 million visitors each year and generate billions of dollars in revenue. Photo credit: T. Fowler.

Vision for the Interagency Oceans and Human Health Research Program

A strong and well-coordinated Interagency Oceans and Human Health (OHH) Research Program will help ensure a healthy ocean where people can swim, fish, recreate, eat seafood and drink water[6] without risk of exposure to dangerous levels of disease-causing organisms, chemical contaminants, or biotoxins and can benefit from ocean-based pharmaceuticals and other natural products.

Photo credit: M. Hollinger.

Value of the Ocean and Role in Enhancing Human Health, Well-Being and Quality of Life

The ocean provides many resources and services that directly benefit human health; seafood is among the most obvious. Seafood supplies high-quality protein, other essential nutrients, and certain healthful omega-3 fatty acids. Numerous studies have shown consumption of fish—especially fatty cold-water species such as salmon, mackerel, sardines, and herring—protects against cardiovascular disease and provides benefits for brain development in fetuses and children. Other studies suggest eating fish may also help protect humans against some cancers, asthma, diabetes, rheumatoid arthritis and other inflammatory diseases, Alzheimer's disease, depression, and macular degeneration. With commercial and recreational fisheries together generating about $80 billion in annual economic activity, plus the increasing importance of aquaculture and dominance of imported products in the U.S. marketplace, maintaining a safe and secure seafood supply is imperative.

The ocean also provides a medicine chest of marine- based pharmaceuticals and other natural products that can greatly enhance our quality of life and combat human illnesses. For thousands of years, terrestrial sources have provided nearly all drug discoveries but with 34 of 36 phyla represented in marine environments, compared to 17 on land, potential discoveries to ben-efit humankind seem boundless. Although only limited effort has yet been expended to discover ocean-source drugs, in part because of the necessity for long-term and high-risk investments, several marine-based products are already on the market, including a pain medication derived from the cone snail, and skin-care additives from the Caribbean sea whip used in cosmetic products. The National Cancer Institute has invested in discovery and development of anti-cancer drugs from the ocean, with over 30 in preclinical and clinical trials to treat a variety of cancers. These marine chemicals are produced by a variety of species, such as bacteria, sponges, ascidians, mollusks, bryozoans, and sharks. Other materials also in trial phases include potential anti-viral and anti-asthmatic drugs from sponges, anti-inflammatory compounds from corals, and treatments for Alzheimer's disease from marine worms. Nutritional supplements, chemical probes for medical research, enzymes, and other useful products are already being derived from the sea. Expanded development of new drugs from the sea is vitally important, but it must be conducted in a sustainable manner.

Wild caught and farm raised seafood have many health benefits such as providing a source of protein, essential nutrients and omega 3-fatty acids for people of all ages. Photo credit: D. Benetti.

The term "seafood" also includes a variety of non-commercial, non-recreational marine life harvested for subsistence purposes by Alaska natives such as walrus and seal meat, seen here drying on a meat rack. Photo credit: G. She iflied.

The oceans provide beneficial pharmaceuticals and products such as the Caribbean sea whip, Pseudopterogorgia elisabethae (Gorgonacae), that contains anti-allergenic and anti- inflammatory compounds used in many products as skin-care additives. Photo credit: V.J. Paul and with permission from The *Oceanography* Society.

Living next to the water provides many benefits such as increased property values, daily aesthetics and general sense of well-being. Photo credit: G. Steele and F. Holland.

Healthy and functioning coastal ecosystems provide clean beaches and drinking water. Much of the coastal economy is powered by tourism, the fastest growing component of coastal business. Over a hundred million Americans and international guests visit our coasts annually, creating billions of dollars in positive eco-nomic impacts and millions of jobs. While tourism creates positive economic returns for coastal residents, they also derive benefits from living next to the water both in terms of property values and in promoting healthy community identity and socio-cultural relationships. This holds true for inland coastal communities as well, such as in the Great Lakes, which provide the additional benefit of supplying drinking water to over 40 million people. Finally, the ocean provides a number of marine species and habitats that can serve as sentinels that may offer early warnings about how the condition of ocean ecosystems may affect humans.

Ocean Health Risks and Associated Economic Impacts

Ocean waters can also negatively impact human health and well-being by exposing humans to seafood-borne or waterborne illnesses that stem from harmful algal blooms (HABs), coastal and industrial pollution, wastewater, sewage, storm water inputs, and point and non-point source runoff from agricultural and animal production systems and other land uses (see Table 1).

For example, seafood may pose risks to consumers and subsistence harvesters because fish, shellfish, and other marine wildlife (e.g., marine mammals and seabird eggs) may accumulate heavy metals (including methyl mercury), persistent organic pollutants (such as PCBs), pesticides, and numerous other chemicals; toxins produced naturally from HABs and other harmful microorganisms; and pathogens (disease-causing organisms). Chemical contaminants can impact human reproduction, cause kidney and brain damage, and have adverse effects on endocrine function, neurobehavioral development, liver function, birth weight, immune responses, and tumor development. Pathogens and toxins sometimes associated with seafood may cause diarrhea, paralysis, amnesia, or death. Contamination of shellfish with pathogens and biotoxins is a continuing problem and every year results in closures of shellfish grounds to harvesting and warnings about shellfish consumption. Even though the FDA's "Fish and Fishery Products Hazards and Controls Guide" and NOAA's National Seafood Inspection Program have improved fishing, handling, and processing protocols to ensure safer seafood products, seafood poisonings are still under-reported and may be increasing in incidence and geographic distribution. For instance, during a period of unusually warm surface waters in 2004, oyster farms in Prince William Sound, Alaska, were devastated by an outbreak of a highly virulent form of *Vibrio parahaemolyticus*, resulting in 63 confirmed human cases plus others in marine mammals.

Shellfish harvest and beach closures are widespread throughout the U.S. due to presence of harmful pathogens (disease-causing organisms) and HAB toxins. Photo credit: L. Younglove.

Table 1. Ocean-Borne Health Threats to Humans.

Disease	Cause	Common Associated Vectors or Exposures	Common Signs and Symptoms
Microbe-related Illnesses			
Viral gastroe- nteritis	Norovirus	Fecal-oral transmission, possibly airborne, ingesti-on of contaminated food or water	Nausea, vomiting, diarrhea, abdominal pain, muscle aches, headache, general sense of tired- ness, low-grade fever
Vibrio parahae- molyticus food poisoning	*Vibrio parahaemolyticus*	Ingestion of raw or inadequately cooked seafood or any food cross-contaminated by handling raw seafood, or by rinsing food with contaminated seawater	Watery diarrhea, abdominal cramps, some- times with nausea, vomiting, fever, and head- ache. Occasionally, a dysentery-like illness is observed, with bloody or mucoid stools, high fever, and high white blood cell count
Vibrio vulnificus food poisoning and/or skin infection	*Vibrio vulnificus*	Food poisoning due to ingestion of raw or inadequa- tely cooked seafood or any food cross-contami- nated by handling raw seafood or by rinsing food with contaminated seawater. Skin infection due to coming into contact with contaminated seawater or raw seafood (e.g., oysters) particularly if open cut on skin	Food poisoning: vomiting, diarrhea, and abdominal pain. Skin infection: gangr ene, sepsis including fever, chills, decreased blood pressure, and possible death (even in healthy people)
Salmonellosis	*Salmonella* sero¯Ingestion types including *S. typhi- murium* and *S. enteritidis*	Ingestion of organisms in food derived from infected food animals or contaminated by feces of an infected animal or person	Acute enterocolitis (sudden onset of headache, abdominal pain, diarrhea, nausea, and vomiting)
Typhoid fever	*Salmonella typhi*	Ingestion of food or water contaminated by feces and urine of patients and carriers, ingestion of raw fish harvested from sewage-contaminated beds, raw fruits and vegetables fertilized with night soil, ingestion of milk contaminated with feces from infected person	Sustained fever, headache, general feeling of tiredness, anorexia, rose spots on the trunk, enlarged spleen, non-productive cough
Shigellosis	*Shigella* species (*S. dysenteriae, S. flexneri, S. boy- dii, S. sonnei*)	Direct or indirect fecal-oral transmission from a pa- tient or carrier, ingestion of water or milk contaminated with feces from infected person	Diarrhea accompanied by fever, nausea, vomit- ing, cramps, and tenesmus (stra-ining at bowel movement or urination)

Table 1. (Continued)

Disease	Cause	Common Associated Vectors or Exposures	Common Signs and Symptoms
Cholera	*Vibrio cholerae*	Ingestion of water contaminated with feces or vomitus of patients, ingestion of unrefrigerated food contaminated with feces from an infected person	Profuse watery stools, occasionally vomiting, rapid dehydration, acidosis and circulatory collapse
Yersinosis	*Yersinia enterocolitica, Yersinia pseudotu-berculosis*	Ingestion of food or water contaminated with feces of infected person or animal	Enterocolitis, pain mimicking appendicitis, fever, headache, pharyngitis, anorexia, vomiting, post-infectious arthritis
Hepatitis	Hepatitis A virus (Enterovirus type 72)	Person-to-person via oral-fecal route, ingestion of food or water contaminated by infected food handlers, ingestion of raw or undercooked mol-lusks harvested from sewage-contaminated beds	Abrupt fever, general feeling of tiredness, an-orexia, nausea, abdominal discomfort, followed by jaundice
Pneumonia (Legionnaire's dis- ease, Pontiac fever)	*Legionellae* species	Airborne via aerosol-producing devices	Anorexia, general feeling of tiredness, muscle weakness, headache, rapidly rising fever, non-productive cough
Respiratory tract infections	Microbes	Inhalation of aerosols of contaminated water	Cough (productive and non), fever
External otitis media	Bacteria & viruses	Contact with contaminated water	"Swimmers ear"
Asceptic (viral) meningitis	Many viruses capable of pro- ducing meningeal features, *Coxsacki-evirus* Group B	Varies depending on virus involved; little evidence of water-borne transmission	Rapid onset of fever with signs of meningeal involvement, increased protein in cerebral-spinal fluid, rash resembling rubella
Cryptosporidiosis	*Cryptosporidium* sp.	Fecal-oral, including person-to-person, animal-to- person, ingestion of water contaminated with feces of infected person or animal	Profuse watery diarrhea preceded by anorexia and vomiting, cramping abdominal pain, general feeling of tiredness, fever
Amoebic dysentery	*Entamoeba histolytica*	Hand-to-mouth transfer of feces, ingestion of water contaminated with amoebic cysts,	Acute dysentery with fever, chills, and bloody or mucoid diarrhea

Table 1. (Continued)

Disease	Cause	Common Associated Vectors or Exposures	Common Signs and Symptoms
		ingestion of contaminated raw vegetables	
Giardia	*Giardia lamblia*	Hand-to-mouth transfer of cysts from the feces of infected individual, ingestion of water contaminated with cysts	Chronic diarrhea, abdominal cramps, bloating, frequent loose and pale greasy stools, fatigue, weight loss
Scombroid poisoning	Histamine-like toxins from bacterial overgrowth	Consumption of spoiled fish, particularly tuna and similar fish	Histamine-like reaction, rash, difficulty breath- ing, vomiting
Anasikiasis	*Anasikiasis* (nematode or round worm) *Gnathostoma* (hel- minth)	Consumption of raw fish (especially cod, herring, mackerel and salmon), and cephalopods (such as squid).	Asymptomatic, eosinophilia, abdominal discomfort, intestinal hemorrhage, granulomatous reaction, allergic reactions
Diphyllobothriasis	*Diphyllobothriasis* (fish tapeworm)	Consumption of raw fish (particularly preparation of gefilte fish but also raw Pacific salmon)	Asymptomatic, megaloblastic anemia (secondary to vitamin B1 2 deficiency), eosinophilia
Skin infection	*Stapholococcus aureus*	Skin contact with contaminated water and/or sand	Skin infection, sepsis, toxic shock syndrome, and death (the latter in very rare cases)
Harmful Algae-related Illnesses			
Amnesiac shell-fish poisoning (ASP)	Domoic acid	Bivalve shellfish, primarily scallops, mussels, clams, oysters & possibly fish	Vomiting, diarrhea, abdominal pain and neurologic effects (confusion, memory loss, disorientation, seizure and coma)
Diarrheic shellfish poisoning (DSP)	Okadaic acid & dinophysistoxins	Bivalve shellfish, primarily scallops, mussels, clams, oysters	Diarrhea, abdominal pain, chills, headache, fever
Neurotoxic shellfish poisoning (NSP)	Brevetoxins	Bivalve shellfish, primarily mussels, oysters, scallops, & possibly fish	Tingling and numbness of lips, tongue, and throat; muscular aches; dizziness; reversal of the sensations of hot and cold; diarrhea and vomiting
Paralytic shellfish poisoning (PSP)	Saxitoxins	Scallops, mussels, clams, oysters, cockles & certain herbivorous fish (e.g., pufferfish), and crabs	Paresthesias of mouth and lips, weakness, dysphasia, dysphonia, respiratory paralysis, diarrhea, nausea, vomiting

Table 1. (Continued)

Disease	Cause	Common Associated Vectors or Exposures	Common Signs and Symptoms
Ciguatera fish poisoning	Ciguatoxins, maitotoxin, scaritoxin	Large reef fish (e.g. grouper, red snapper, amberjack, barracuda [most common])	Abdominal pain, nausea, vomiting, diarrhea followed by paresthesias, reversal of hot/cold, pain, weakness. Can proceed to bradycardia, hypotension, increase in T wave abnormalities. Symptoms may persist for up to years.
Azaspiracid shellfish poisoning (AZP)	Azaspiracid	Bivalve shellfish	Nausea, vomiting, liver damage
Florida red tide respiratory irritation	Brevetoxins	Aerosolized seawater	Acute eye irritation, acute respiratory distress (nonproductive cough, rhinorrea), asthma exacerbation
Blue green algae (Cyanobacteria)	Anatoxins, axitoxins, icrocystins, nodula- rins, cylindrosper- mopsins	Direct contact with water, inhaling aerosols, drink ing water, and possibly seafood	Skin and eye irritation, nausea, vomiting, respiratory distress, liver damage
Marine Creatures			
Fugu and pufferfish poisoning	Saxitoxin, Tetrodotoxin	Consumption of certain species of pufferfish	Paresthesias of mouth and lips, weakness, dysphasia, dysphonia, respiratory paralysis, diarrhea, nausea, vomiting
"Sea bather's eruption" ("swimmer's itch" or "sea lice")	Larvae of jellyfish such as *Linuche unguiculata* (Thimble jellyfish); and *Schistosoma cercaria* (free-swimming flatworm found in Florida)	Skin contact	Itchy rash, hives, particularly along bathing suit lines; may be more severe in children; allergic reactions
Jellyfish stings	Jellyfish such as: *Cyanea capillata* (Lion's mane jellyfish), *Aurelia aurita* (Moon jelly), *Stomolophus meleagus* (Cann-	Skin contact with jellyfish, particularly the tentacles	Burning itchy rash, hives, rarely systemic reaction and even death; allergic reactions

Table 1. (Continued)

Disease	Cause	Common Associated Vectors or Exposures	Common Signs and Symptoms
	onball jellyfish), *Cassiopeia xamachana* (Mangrove upside-down jellyfish), *Pelagia noctiluca* (Purple jellyfish), *Chrysaora quinquecirrh*a (Sea nettle), *Carybdea alata* (Sea wasp), *Linuche unguiculat*a (Thimble jellyfish), *Drymonema dalmatinum* (Stinging cauliflower)		
Fire coral	Fire Coral such as: *Millepora complanata* (Blade fire coral), *Millepora alcicornis* (Branching fire coral), *Millepora squarrosa* (Box fire coral), *Tedania ignis* (Fire sponge), *Microciona prolifera* (Red moss sponge), *Fibulia nolitangere* (Poison bun sponge)	Skin contact with fire coral	Rash, itching, burning, allergic reactions
Sea urchin spines	Sea urchins such as: *Diadema antillarum* (Long-spined sea urchin), *Diadema setosum* (Black spiney sea urchin)	Penetration of skin by spine(s)	Skin puncture, pain, tenderness, secondary bacterial infections at site of puncture

Table 1. (Continued)

Disease	Cause	Common Associated Vectors or Exposures	Common Signs and Symptoms
Sting ray spines on tail	Sting ray	Penetration of skin by spine on tail	Skin puncture, pain, tenderness, secondary bacterial infections at site of puncture; allergic reactions; rarely death
Lionfish and scorpionfish spines	Lionfish and Scorpion-fish such as: *Scorpaena plumieri* (Spotted scorpionfish), *Scorpaena grandiconis* (Plumed scorpionfish), *Scorp-aenna brasiliensis* (Barbfish), *Pterois volitans* (Red firefish /Lionfish)	Penetration of skin by spine(s)	Pain, headache, nausea, confusion, fainting, death
Catfish spines	Catfish such as: *Arius felis* (Hardhead catfish), *Bagre marinus* (Gafft-opsail catfish), G*aleich-thys felis* (Common saltwater catfish)	Penetration of skin by spine(s)	Pain, swelling, numbness; rarely, fainting, reduced heart rate
Bites	Sharks and other predatory species such as wahoo, barracuda and seasnakes	Penetration of skin, muscle, bone, viscera by teeth	Pain, amputation, death

Humans can also be exposed to pathogens, biotoxins, or chemical contaminants by swimming or recreating in ocean waters or drinking tainted water. The CDC, EPA, and the Council of State and Territorial Epidemiologists maintain a collaborative surveillance system for collecting and reporting on the occurrence and cause of waterborne disease and outbreaks in drinking and recreational waters. However, outbreaks resulting from exposures to recreational waters, particularly marine waters, are often under-reported and/or misdiagnosed. Further, CDC concluded that the incidence of infections associated with recreational water use has steadily increased over the last several decades. Estimated increases in occurrence of gastrointestinal illness caused by swimming in contaminated waters at beaches in Orange and Los Angeles Counties, California correspond to an annual negative economic impact of $21 to $51 million. If one extrapolates from these results for just two counties to the more than 20,000 instances of closings and advisories annu-ally at coastal and Great Lakes beaches, the public health costs associated with illnesses potentially occurring as a result of exposures on beaches could be staggering.

HABs are proliferations of microscopic algae that harm the environment and humans by producing toxins that accumulate in shellfish and fish and concentrate up the food chain. Such blooms occur naturally but have increased in number, frequency, and type in recent years. The biotoxins produced by HABs are not destroyed by cooking or other methods of food preparation and can cause serious illness and even death in humans. HABs have been a persistent problem in several areas of the country, including Alaska, with some documented human deaths caused by paralytic shellfish poisoning from consumption of contaminated shellfish. Other HAB toxins, such as domoic acid, could become more prevalent as a result of climate change. People can also be affected by inhalation of aerosolized HAB toxins while walking on the beach or from coming into direct contact with contaminated waters. Recent work by the CDC and OHH researchers found significant increases in reported respiratory symptoms among recreation- al beachgoers in Florida after a red tide (*Karenia* spp.) exposure. Another study showed emergency room visits related to respiratory diagnoses rose 54 percent in years when red tides occurred in coastal Florida. While probably significantly underestimated, the overall economic effects of HABs in the U.S. (not including freshwater outbreaks) are at least $82 million per year, with public health costs of illness accounting for 45 percent of the total, commercial fisheries impacts for 46 percent, local recreation and tourism impacts for 5 percent, and costs of coastal monitoring and management for 4 percent.

A red tide (*Karenia brevis*) bloom along a Florida beach impacts local tourism and can induce respiratory symptoms among beachgoers. Photo credit: P. Schmidt and with permission from TOS.

Other ocean health threats include the prevalence of emerging chemical contaminants (ECC), such as drugs used to treat diseases in humans and domestic animals, flame retardants, stain repellants and non-stick coatings, and pesticides in aquatic environments. Numerous such compounds are rapidly accumulating in coastal waters, appear to be persistent, and some are concentrated as they move up the food chain, as has been documented in marine mammals that feed at a similar trophic level as humans. Possible effects on estuarine organisms include cancer; impacts to reproductive physiology, early development, immune function, and metabolism; and other species-specific effects that could impair healthy ecosystem functioning. The health effects of human exposures to these chemicals via the complex circulation of environmental contaminants through air and water are not well understood and a coordinated approach to assessing the effects of these exposures is needed.

Coastal population densities are two–five times higher than inland, and coastal land is being consumed for urban development three–six times as fast as the rate of population growth, resulting in major and permanent alterations to coastal ecosystems. These trends appear to be accelerating, with potentially serious impacts on the long-term health of coastal ecosystems and the people who live, work and recreate there. The latest assessments of coastal-ecosystem condition can be found in the National Coastal Condition Report II (NCCR) prepared by the EPA with contributions from NOAA, the U.S. Fish and Wildlife Service (USFWS), and the USGS. The report, using data from the period 1997–2000, rated U.S. coastal waters as generally fair, with some poor areas, in terms of water quality, sediment contamination, and status of biological resources. Relatively few areas were rated good, and little improvement was noted over the previous assessments in NCCR I, which used data from 1990–1996.

Newly recognized emerging chemical contaminants such as flame retardants, used in furniture and electronics, are rapidly increasing in coastal waters, bio-magnify up the food chain and may affect the health of aquatic species and people.

U.S. coastlines are highly desired locations for residents and visitors with rates of urban development much higher than the national average, resulting in significant alterations to coastlines and impacts to coastal ecosystem resiliency. Photo credit: R. Kresge.

Brief History of the Interagency Oceans and Human Health Program

Widespread recognition that the ocean plays an important and growing role in affecting human health was sparked when the U.S. sponsored an OHH-themed pavilion at the 1998 World's Fair in Lisbon, Portugal, followed soon thereafter by publication of the National Research Council's (NRC) 1999 seminal report, *From Monsoons to Microbes: Understanding the Ocean's Role in Human Health.* Subsequent NRC reports, such as *Under the Weather: Climate,Ecology and Infectious Disease* (2001) and *Marine Biotechnology in the Twenty-First Century: Problems, Promise, and Products* (2002), and workshops led by NSF, NIEHS, and NOAA further substantiated the need for action at the federal level. Together, these activities and reports highlighted the importance of OHH activities and coordination among federal agencies.

NSF and NIEHS joined forces, held a community workshop in December 2001, and issued a Request for Proposals in November 2002 for the establishment of academic centers for OHH. For NOAA, Congress appropriated funding beginning in FY 2003 for the agency to establish an Oceans and Human Health Initiative (OHHI) to "coordinate and focus agency activities on critical areas of concern and identify critical gaps in coverage, and ... to be used for critical research and projects aimed at closing identified gaps." Congress directed NOAA to consult with NSF and other federal agencies, establish NOAA Centers of Excellence in OHH, create a national advisory panel, and support competitive external grants, distinguished scholars, and trainees in OHH. NOAA sponsored a community workshop in October 2003 to solicit input for development of its OHHI.

The NSF and NIEHS jointly awarded five-year OHH Center grants to four academic institutions in May 2004. NOAA selected its three OHH Centers of Excellence via an internal peer-reviewed competition a few months later, followed by initiation of its external OHHI grant program. Also in 2004, the USCOP released its final report, which spoke to the importance of a national OHH program. In December 2004, Congress passed the OHH Act (see Appendix 1) and established an Interagency OHH Research Program to "... improve understanding of the role of the oceans in human health" and officially authorized NOAA's OHHI. The Administration's *Ocean Action Plan* response to the USCOP report also included a commitment to "...develop a strategic research plan for oceans and human health."

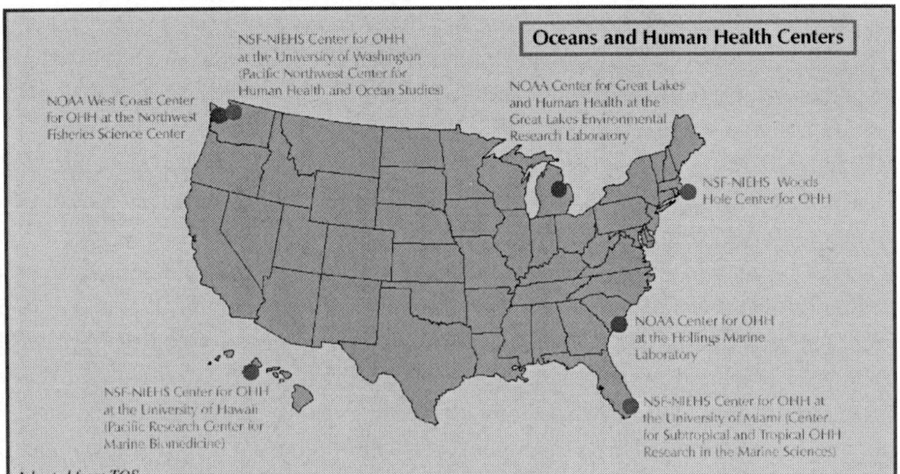

Adapted from TOS.

Requirement for an Interagency OHH Research Implementation Plan and Coordination with the National Ocean Research Priorities Plan

This Interagency OHH Research Implementation Plan was prepared pursuant to requirements of the OHH Act of 2004. The Plan builds upon the institutionalized and formally named "OHH" programs of NOAA, NSF, and NIEHS and the expressed interest of other agencies with a wealth of OHH-related activities, including the CDC, EPA, FDA, MMC, NASA, and USGS. As stated in the OHH Act, the purpose of the Implementation Plan is to establish "the goals and priorities for Federal research which most effectively advance scientific understanding of the connections between the oceans and human health, provide usable information for the prediction of marine-related, public health problems, and use the biological potential of the oceans for development of new treatments of human diseases and a greater understanding of human biology." The OHH Act also directed the National Ocean Research Leadership Council to ensure that ocean and coastal observations provide useful information necessary to reduce marine public health problems.

This Interagency OHH Research Implementation Plan was developed by the Interagency Working Group on Harmful Algal Blooms, Hypoxia, and Human Health (IWG-4H). In fall 2005, the Joint Subcommittee on Ocean Science and Technology (JSOST) of the National Science and Technology Council (NSTC) Committee on Science and the Committee on Environment and Natural Resources and the Interagency Committee on Ocean Science and Resource Management Integration (ICOSRMI) established the IWG-4H. The ICOSRMI and JSOST are planning and coordinating bodies under the President's cabinet-level Committee on Ocean Policy. This document presents a comprehensive interagency effort to coordinate and enhance OHH-relat-ed research over the next decade. It complements and expands upon the "Enhancing Human Health" theme described in *"Charting the Course for Ocean Science in the United States for the Next Decade: An Ocean Research Priorities Plan and Implementation Strategy"*, prepared by the JSOST pursuant to the Administration's *Ocean Action Plan*. Numerous participants in the Interagency OHH Program contributed to the development of the "Enhancing Human Health" theme, which highlights four OHH-related research priorities:

- Understand sources and processes contributing to ocean-related risks to human health.
- Understand human health risks associated with the ocean and the potential benefits of ocean resources to human health.
- Understand how human use and valuation of ocean resources can be affected by ocean-borne human health threats and how human activities can influence these threats.
- Apply understanding of ocean ecosystems and biodiversity to develop products and biological models to enhance human well-being.

"Charting the Course for Ocean Science" also identifies four near-term priorities to be pursued over the next two to five years to initiate rapid progress toward the 20 ocean research priorities highlighted in the plan. One of the near-term priorities is the development of new sensors that will, among other things, provide essential information to enable forecasting of ocean-related risks to human health and safety, including those associated with pathogens, HABs, extreme weather events, natural disasters, and changing ocean conditions.

Goals for the Interagency OHH Research Implementation Plan

This plan focuses on coordinated federal activities that will foster "interdisciplinary research among the ocean and medical sciences, and coordinated research and activities to improve understanding of processes within the ocean that may affect human health and exploration of the potential contribution of marine organisms to medicine and research," as required by the OHH Act of 2004. The Interagency OHH Program will focus on the following broad goals:

Establish an OHH research agenda focused on the highest priority topics and activities - including the appropriate mix of infrastructure to allow scientists to develop a requisite understanding of ocean-system properties and processes; assess, predict, and prevent human health risks and impacts; and discover and develop new marine bio-products that benefit mankind. This research agenda must also incorporate economic and socio-cultural dimensions, as well as epidemiologic approaches, into a fuller understanding of ocean-human health linkages. It should capitalize on further development and application of advances in such areas as genomics, proteomics, metabolomics, analytical chemistry, microbiology, source tracking, aquaculture, use of biological sentinels and indicators, and other new or improved approaches for tracking movements of pathogens, toxins, and chemical contaminants through the environment.

Promote a robust interdisciplinary OHH research approach - that links marine and biomedical sciences to create a fundamentally new and dynamic research community with the breadth of expertise required to address diverse human health questions within the complexity of ocean systems

Establish and expand collaborative partnerships - that link academic, federal, state, and private researchers with public health workers, environmental/natural resource managers, and multiple user groups to facilitate the acquisition of OHH information and its application through technology transfer and training. Such partnerships will ensure inclusion of OHH-supportive data collection, integration, analysis, and modeling within the larger framework of national and international ocean observation efforts.

Expected Benefits to the Nation from OHH Research

The interagency OHH planning effort has identified seven major benefits that will accrue to the United States as a result of continued and coordinated research in OHH, accompanied by rapid application of research findings to real-life situations involving health threats to humans and potential for development of health-enhancing products.

1) **Improved policy decisions that reduce health risks and help ensure healthy and productive marine ecosystems and people via enhanced monitoring and health surveillance for humans and wildlife** - These activities include developing and employing a range of biosensors, models, databases, and predictive tools to provide early warning and other information for management, control, and prevention by resource and public health managers and policy-makers. Connecting data from new biosensors for pathogens, contaminants, HABs, and toxins with data from ocean and coastal observing systems and new epidemiological studies of human disease will enable significant advances in protecting the public from ocean- borne health threats. In addition, implementation of sophisticated techniques for surveillance of disease in ocean wildlife populations will substantially decrease the risk to humans from diseases that may jump from animals to humans, such as avian influenza.

2) **Discovery and application of marine-based pharmaceuticals and other beneficial natural products** - A broad effort to discover, test, develop, and market new products, including drugs, from the ocean will result in improved human health, enhanced private sector returns, increased employment opportunities, and development of a new ocean business sector based on renewable resources and novel ecologies and chemistries.

3) **Increased basic knowledge** - A strong Interagency OHH Research Program will produce significant advances in basic scientific information and thereby provide a stronger foundation for understanding and preventing health threats to humans and managing coastal development and land to minimize adverse environmental and health impacts on people and coastal ecosystems.

4) **Enhanced understanding of human disease processes** - This program will develop and utilize new mechanistic and conceptual models derived from marine species and marine sentinel organisms and habitats. These models will enable major advances in

understanding a broad range of human-disease processes, resulting in improvements in prevention, intervention, and treatments to protect human health.

5) **Improved understanding and communication about comparative benefits and risks of seafood consumption** - Coordinated interagency work will allow the nation to maximize health benefits and minimize risks from the consumption of seafood and will provide essential information to empower consumers and subsistence users to make well- informed individual choices about seafood.

6) **Increased economic returns** - Application of OHH research results is expected to lead to decreased economic losses related to beach closures, lowered risks for exposure of people to health risks from ocean waters and related health care costs, and maintenance of coastal property values through enhanced ability to minimize occurrences of ocean health threats. In addition, major new business development in a burgeoning ocean natural products sector should lead to increased economic returns.

7) **Improved ocean stewardship and ocean literacy** - Effective communication of the effects of the oceans on human health and vice versa will enable a well-informed public to act as stewards of coastal and marine ecosystems, thereby reducing or preventing human health risks and maximizing human health benefits.

Coordination of oceanographic research across agencies, such as at-sea sampling, will provide many benefits to the U.S. Photo credit: NOAA Northwest Fisheries Science Center (NWFSC).

2: Existing Agency OHH Programs and Related Activities

The OHH Act of 2004 established "...a national research program to improve understanding of the role of the oceans in human health" and stated that such a program "will build on and complement the ongoing activities of the National Oceanic and Atmospheric Administration, the National Science Foundation, and other departments and agencies." NOAA, NSF, and NIEHS have specifically authorized research activities focused on OHH, and several other agencies have missions related to OHH issues (e.g., CDC, EPA, FDA, MMC, NASA, USGS). Current OHH activities of individual agencies, as well as some important interagency collaborations, are summarized in this chapter. Excellent examples of interagency collaboration include the OHH program's responses to Hurricane Katrina, the Massachusetts Bay HAB event, and epidemiologic studies of health effects of aerosolized Florida red tide toxin, as described herein. More information on interagency OHH program activities will be presented in the OHH Annual Report, to be published in the near future.

Named OHH Programs (NSF, NIEHS and NOAA)

NSF and NIEHS

In June 2000, NSF and NIEHS began discussions about a collaborative program with the goal of bringing together the expertise of the medical community and the ocean science community to create "Centers of Excellence" where OHH problems would be addressed in an interdisciplinary research context (see Appendix 2: Agency Prospectus for NSF). HABs, water- and vector-borne diseases, and the search for marine pharmaceuticals were identified as foci for the initial OHH effort. In December 2001, the agencies convened a workshop involving 24 scientists, representing a broad range of relevant expertise, to outline the scientific issues and discuss strate-gies for implementation of an interdisciplinary initiative. NSF and NIEHS used recommendations from this workshop to develop an Interagency Request for Applications (RFA) to fund several OHH Centers. These agencies received proposals in March 2003 and, after subsequent peer review, they funded four centers in 2004.

NSF and NIEHS expressed an overarching vision for their OHH Centers in the call for applications:

> *"The field of oceans and human health currently represents a gap between the two participating agencies missions. As such, it also represents a logical opportunity for partnership. Together, NSF and NIEHS plan to bridge this gap and promote state-of-the-art, interdisciplinary research that unites the oceanographic and medical communities, allows for cross-fertilization of ideas and technologies, and provides more comprehensive insight of the potential risks and benefits to human health generated by the oceans."*

The four NSF-NIEHS centers are in their fourth year of operation and have collaborated with a number of non-affiliated academic institutions, formed working partnerships with several federal agencies (CDC, EPA, USGS, USDA, Naval Research Lab [NRL], NOAA) and state and local heath departments, grassroots groups, and others. To date, Center activities have involved more than 60 senior researchers, more than 10 postdoctoral researchers, and more than 40 graduate students and undergraduates. Already, at least 40 publications have resulted from the Centers' research activities, and there have been many presentations of OHH activities at national and international scientific meetings. The combined investment for NSF and NIEHS for the present five-year duration of the Centers' awards is approximately $25M.

A brief overview of Center projects, partnerships, some initial successes, and education and outreach efforts is provided here.

NSF-NIEHS Woods Hole Center for Oceans and Human Health (WHCOHH)

The mission of the WHCOHH is to improve public health through enhanced understanding of how oceanic processes affect the distribution and persistence of human pathogens and toxin-producing organisms. The Center is a partnership that includes the Woods Hole Oceanographic Institution (WHOI), the Marine Biological Laboratory, and the Massachusetts Institute of Technology. WHCOHH researchers study the distribution of biological agents in temperate coastal ocean waters that may negatively impact human health.

Two research projects focus on blooms of toxic dinoflagellates of the genus *Alexandrium*. *A. fundyense* is responsible for paralytic shellfish poisoning (PSP), and *A. ostenfeldii* is associated with an emerging toxin threat. One project explores the extent of natural genetic diversity within *Alexandrium* in the Gulf of Maine and will develop genetic markers capable of distinguishing among the different *Alexandrium* genotypes. A second coordinated project is developing a suite of population- dynamics models to explain or predict the distribution of *Alexandrium* genotypes within the Gulf of Maine. These HAB projects build on previous studies funded by NSF and NOAA. Results from the WHCOHH HAB-focused projects will provide a comprehensive understanding of toxic *Alexandrium* bloom dynamics which is required to model and predict human exposure to toxin through harvested shellfish. One of the initial successes of the combined HAB projects was the mapping of a massive red tide that occurred in Massachusetts Bay in May of 2005 (see Box 1).

The two other WHCOHH projects address the distribution of pathogens in coastal waters. The first examines the diversity, distribution, and persistence of human pathogens in the coastal marine environment of Mt. Hope Bay, Mass., an estuarine environment heavily impacted by humans. These initial studies will assist in establishing and validating a water quality model for Mt. Hope Bay and a model system for identifying and preventing human health problems from water-borne pathogens in other coastal environments (see Box 13). The second project examines the ecology and evolution of two important human bacterial pathogens, *Vibrio vulnificus* and *V. parahaemolyticus*, in the Plum Island estuary, Mass.. This project is expected to yield major insights into what determines the presence of pathogens in coastal waters and what environmental factors select for the presence of pathogenicity traits in bacteria. These in- depth studies of pathogens are among the first to examine the dynamics and diversity of human pathogens in coastal waters on multiple spatial and temporal scales.

Pilot projects are augmenting ongoing research and extending research in the areas of natural toxins and natural bioactive compounds. A genomics core facility at WHCOHH provides DNA-sequencing services, microsatellite analysis, and computational support to Center investigators and collaborators.

Researchers at WHCOHH are cooperating with the USGS, NRL, and USDA. HAB research has included international collaborations with the National Research Institute of Fisheries and Environment of the Inland Sea in Hiroshima, Japan, and participation by a WHOI researcher in a new European Commission-U.S. collaborative project.

Box 1: Massive Red Tide in Massachusetts Bay

In May 2005, with funding and other support provided by NOAA and NSF and in collaboration with other agencies (see Box 45), researchers at the WHCOHH applied cutting-edge marine field instrumentation and biochemical technology to study and map the largest outbreak of red tide in Massachusetts Bay in more than 30 years. The research team used in-water sensors (including those operated by the Gulf of Maine Ocean Observing System), state-of-the art shipboard optical and molecular biological techniques, and coupled ecological-physical models to characterize the initiation and development of a bloom of *Alexandrium fundyense*, the major red tide species in the Gulf of Maine. These combined capabilities, organized under the umbrella of the WHCOHH, enabled the researchers to characterize the development of this toxic bloom in real time. Scientific understanding of what factors initiate red tides in the region is benefiting resource managers charged with monitoring shellfish for threats to human health. The team's discovery and tracking of this bloom helped prevent serious public health consequences.

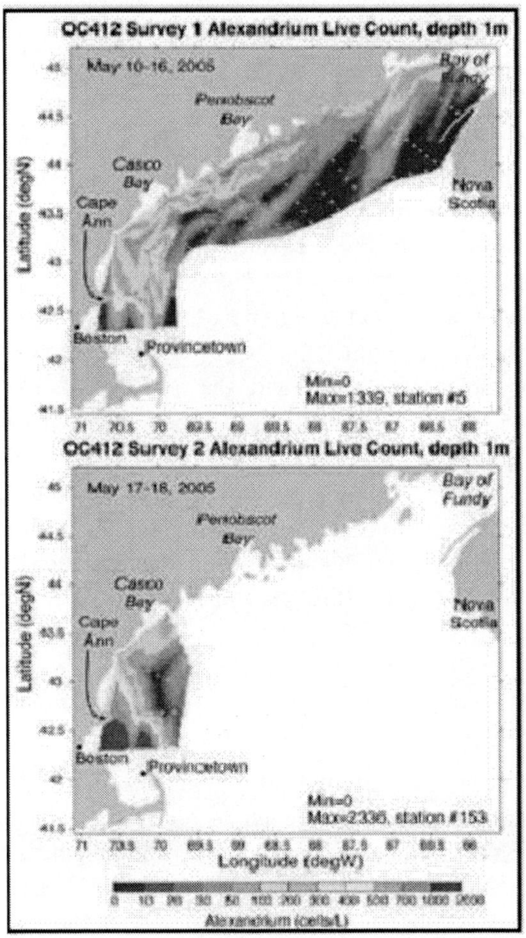

Alexandrium concentration mapped from shipboard measurements.

NSF-NIEHS Center for Oceans and Human Health at the University of Miami (Center for Subtropical and Tropical Oceans and Human Health Research in the Marine Sciences [UMCOHH])

The primary mission of the UMCOHH is to focus innovative interdisciplinary scientific research on OHH issues affecting the subtropical and tropical oceans. The Center's partners include Florida International University, Florida Department of Health, University of Florida, Nova Southeastern University, CDC, University of North Carolina at Wilmington, Mote Marine Laboratory, Florida Poison Information Center at Miami, Lovelace Respiratory Institute, and NOAA.

One important study is investigating the use of microbial indicators of human health effects in subtropical beaches and other recreational marine waters. This research incorporates modeling of the physical oceanographic environment with the characteristics of microbes causing or indicating public health problems on beaches. The goal of this project is to discover basic, predictive indicators of coastal contamination, including residence times and transport pathways of pathogenic and indicator microbes in coastal systems.

A second research project focuses on discovery and characterization of new HABs. Many species of dinoflagellates, cyanobacteria, and other algae produce toxins that are not presently noticed as harmful to humans because those species generally do not occur in dense-enough concentrations to produce harmful quantities of toxins or because they are in areas of the ocean where humans are not exposed to the toxins. However, circumstances are likely to change, with the possibility that these previously unrecognized toxin producers could become important threats to human health (see Box 2). Another HAB-related project is using a functional genomics approach to examine toxin production by the Florida red tide organism, *Karenia brevis* (see Box 33).

The CDC and NOAA are organizational partners at the University of Miami OHH Center. UMCOHH personnel also collaborate with NIEHS researchers at NIEHS Marine and Freshwater Biomedical Sciences Center and the NIEHS -funded Florida International University-University of Miami Advanced Cooperation in Environmental Health Research Program. In addition, the UMCOHH collaborates with the University of North Carolina, Wilmington, the CDC, Florida Department of Health, Mote Marine Laboratory, Lovelace Respiratory Institute, the Florida Marine Research Institute and other partners in an ongoing NIEHS-funded project to investigate human and animal exposures to and health effects of aerosolized Florida red tide toxin (see Box. 25).

NSF-NIEHS Center for OHH at the University of Hawaii (Pacific Research Center for Marine Biomedicine, PRCMB)

PRCMB builds on the University of Hawaii's research strengths in *Oceanography* and tropical medicine and its location in the center of the largest ocean on Earth. Through the collaboration of an interactive milieu of oceanographers and medical researchers, the PRCMB is conducting hypothesis-driven, interdisciplinary research on HABs, water- and vector-borne diseases, and marine- derived pharmaceuticals and probes, in the thematic context of tropical coastal waters and small islands. In conducting its programs, PRCMB is partnering with other universities; federal, state, and local agencies; and the private sector.

BOX 2: DISCOVERY OF NEW TOXIC ALGAL SPECIES

Over the past few decades, multiple species of harmful algal organisms with newly discovered toxins have appeared, possibly due to increased nutrient loading in coastal waters. Many other species of marine algae also may produce toxins, but they have not yet been recognized because they do not occur in high-enough concentrations or are not located where they can cause problems to humans and other animals at present. However, it is highly likely that some of these algae will become HABs in the future, and that low concentrations of their toxins already occur in ocean food webs and ecosystems.

The UMCOHH is examining numerous algal species from subtropical and tropical waters for the production of a variety of types of HAB toxins. For this exploratory work, the Center has developed automated methods for isolating viable single cells using flow cytometry. To date, researchers have isolated 67 uni-algal cultures. Genomic sequences indicate that many of these isolated strains are previously undescribed HAB species. Preliminary screening has already indicated that at least 22 of these isolates, representing an as-yet unknown number of species, are potentially toxic to humans and other organisms.

Researchers at UMCOHH maintain subtropical and tropical algal cultures to screen for existing and emerging HAB species. Photo credit: K. Rein.

The PRCMB's HAB-related work is focused on ciguatera fish poisoning, attempting to develop methods to detect ciguatera poisons and determine what is responsible for the variability of ciguatera toxins in nature (see Box 22). Another research project is studying local water qual-ity and human health effects from water-borne pathogens. Specifically, this project is attempting to identify useful indicator organisms for sewage contamination of Hawaiian (tropical) coastal waters, developing a procedure to use mollusks to monitor water for viruses, and monitoring beach environments for staph bacteria (*Staphylococcus aureus*) virulence genes to determine what portion of the bacterial population represents a human health risk. The health effects of these pathogens in coastal waters are being determined based on the numbers of virulence characteristics they exhibit and the exposure and susceptibility of human populations. A third research project deals with exploring the ocean for sources of new pharmaceuticals and biochemical probes and discovering new, biologically active natural prod-ucts. The Pharmaceutical Lead and Probe Discovery Program and the PRCMB Core Facility for the Culture and Characterization of Marine Microbes are working closely on the discovery of novel bioactive compounds. This process begins by taking samples from Hawaii's various marine environments, ranging from the nutrient- poor waters of the open ocean to the nutrient-rich environments of Kaneohe Bay. The PRCMB core facility is using innovative, high-throughput culturing techniques to generate pure cultures of heretofore unculturable marine microbes and is providing marine microbe culture and characterization support for all the Center's research projects. Screening of cyanobacterial cultures has led to the discovery of the widespread occurrence of a neurotoxic amino acid (see Box 3).

Box 3: Cyanobacterial Neurotoxin Discovered at PRCMB

In April 2005, researchers at the University of Hawaii's Center for OHH, in conjunction with colleagues at the Institute for EthnoMedicine and elsewhere, reported that a neurotoxin, ß-N-methylamino-L-alanine (BMAA), is produced by all known groups of cyanobacteria in freshwater, marine, and terrestrial environments. Previous investigations of this neurotoxin in the cyanobacterial symbionts of cycad palm trees in Guam led researchers to hypothesize that BMAA is the probable cause of the high incidence of amyotrophic lateral sclerosis/parkinsonism dementia complex (ALS/PDC) among the Chamorro people. In Guam, human exposure to high quantities of BMAA results from unique components of the traditional Chamorro diet, including cycad tortillas, flying foxes, and possibly other feral animals. The ubiquity of cyanobacteria in diverse terrestrial and aquatic environments suggests that ingestion of BMAA, with consequent potential health threats, may occur through even less esoteric routes, including direct consumption of cyanobacteria or cyanobacterial hosts, bioaccumulation in food chains, or exposure to cyanobacteria-contaminated water supplies.

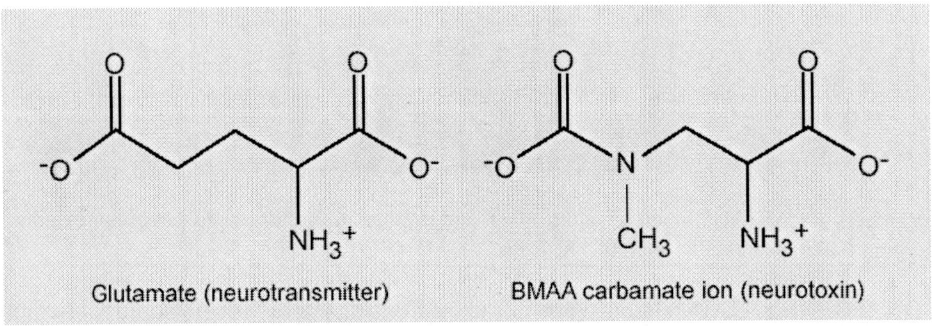

BMAA chemical structure. High exposures to BMAA may be linked to the occurrence of Lou Gehrig's and Parkinson's disease.

At PCRMB, the ciguatera HAB project collaborates with the NOAA marine toxin facility in South Carolina in characterizing the ciguatera toxins. There is also close coordination with the Hawaii Department of Health and the Hawaii Aquaculture Development Program to share information about ciguatera-toxin problems in Hawaii waters. This project also provides algal material to the Cancer Research Center of Hawaii, an NIH/ National Cancer Institute facility at the University of Hawaii, to screen for interesting bioactive compounds. The Center's water-borne pathogen project has collaborated with the Division of Environmental Quality, the Watershed Division of the City and County of Honolulu, the Hawaii Department of Health, CDC, EPA, Staub Clinic and Hospital, Queens Medical Center, and the Surfrider Foundation. The Pharmaceutical Lead and Pharmacological Probe Discovery research project has formed a working partnerships involving NOAA's Center for Oceans and Human Health at the Hollings Marine Laboratory and with Diversa Corporation, a biotechnology company based in San Diego, CA (see Box 23).

NSF-NIEHS Center for OHH at the University of Washington (Pacific Northwest Center for Human Health and Ocean Studies, PNW H2O)

The goal of the PNW H2O is to elucidate relationships between marine processes and public health consequences of HABs, by focusing specifically on the toxic diatom *Pseudo-nitzschia*. Studies are being conducted on *Pseudo-nitzschia* dynamics and human impacts in two major representative ecosystems, the Puget Sound estuary and Washington's Pacific coast. The Center is exploring underlying differences between the coastal environments in the Pacific Northwest and Puget Sound. The Center is composed of a multidisciplinary research team that includes researchers from universities as well as state and federal government agencies, with organizational partners that include the University of Washington, Institute for Systems Biology, Washington State Department of Ecology, Washington State Department of Health, and NOAA's Northwest Fisheries Science Center and its West Coast Center for Oceans and Human Health (WCCOHH).

One research project is focused on the physical, chemical, and biological factors that promote blooms and toxin release from *Pseudo-nitzschia* cells by coupling molecular and genomic techniques with environmental data. This project's objectives are to identify nearshore and offshore environmental factors that influence the dynamics and toxicity of *Pseudo-nitzschia* blooms (see Box 4). This project is using genomic techniques to examine the genetic variations among *Pseudo-nitzschia* populations and to determine the underlying mechanisms of toxin production (see Box 34). This project builds on results obtained from previous funding from NSF through the Ecology and *Oceanography* of HABs (ECOHAB) and the Biological *Oceanography* Program.

A second PNW H2O project seeks to develop a predictive understanding of the factors that regulate the uptake and accumulation of toxic domoic acid in benthic bivalve mollusks. Researchers are attempting to model the transport of toxic algae from offshore waters to nearshore benthic organisms and determine the rate of accumulation and retention of the toxin in local benthic organisms, with special emphasis on commercially harvested species (see Box 20).

A third research project examines the specific molecular mechanisms that account for the toxicity of domoic acid in humans and other vertebrates. This project studies how age affects susceptibility to this neurotoxin and the health risks posed by low-level chronic exposure to domoic acid (see Box 30).

The fourth project focuses on human exposure to domoic acid. Project investigators are partnerning with regional populations to better understand dietary behaviors that contribute to increased exposure and to understand factors that may define "at risk" populations. Such factors can include concurrent dietary exposures, age, cultural practices, and geographic proximity to contaminated shellfish. The human exposure studies provide the link between the Center's basic research on mechanisms of toxicity and environmental domoic acid contamination (see Box 24).

BOX 4: DETECTING THE TOXIC DIATOM *PSEUDO-NITZSCHIA* IN PUGET SOUND WATERS

Pseudo-nitzschia species differ in the amount of toxin they can produce, yet differentiating species in a seawater sample is difficult or impossible with traditional microscopy. To predict toxic events, it is essential to be able to rapidly identify blooming species. Using a DNA- based technique known as ARISA, the species in a sample can now be "fingerprinted" and easily identified.

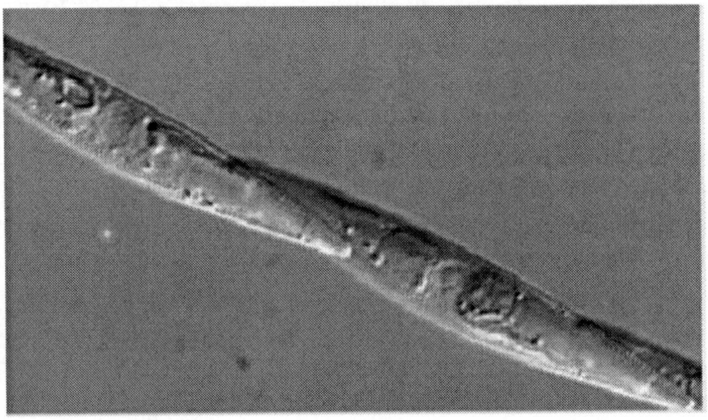

Pseudo-nitzschia diatoms produce high concentrations of domoic acid during algal blooms which can poison marine mammals and humans who eat contaminated shellfish. Photo credit: P. von Dassow.

Resources include real-time environmental monitoring, a biosensor development core facility (joint with the university's Genome Sciences), a neurobehavioral facilities core that conducts animal neurobehavioral studies, and a bioinformatics core that provides statistical and other bioinformatic support critical for translation of findings across disciplines.

Outreach, Education, and Training at the NSF-NIEHS COHH

Outreach

The NSF-NIEHS OHH Centers undertake a number of outreach efforts:

- They maintain public Web sites that can be accessed individually or through a link at WHCOHH, http://www.whoi.edu/science index.htm.
- They have regular seminar programs where OHH-related topics are presented. These are generally targeted to the local academic community, but most are open to the interested public.
- They have developed outreach materials in collaboration with local Centers for Ocean Sciences Education Excellence (COSEE).

Some Center activities are specifically directed toward local outreach. For example, one pilot project at the WHCOHH provides information about northeast HABs to the public and stakeholders in response to a bloom. The PNW H2O at the University of Washington conducts outreach efforts to strengthen partnerships with local Native American Tribes and Asian Pacific Islander communities.

Education and Training

The NSF-NIEHS OHH Centers provide educational opportunities and training to postdoctoral researchers, graduate students, and undergraduates as an inte-gral part of the awards made to the Centers. Additional undergraduate participation at the Centers is encouraged with Research Experiences for Undergraduate Students ([REU]; see Box 5) supplemental awards to all Centers. Already, over ten postdoctoral researchers and more than 40 graduate students and undergraduates have been involved in Center research activities during their first three years of operation. Postdoctoral researchers and students receive training and experience in a number of fields, including physical, chemical, and biological *Oceanography*; remote sensing; genomic techniques; bioinformatics; epidemiology; toxicology; marine biology; fisheries; statistics; and microbial ecology. In addition to formal education, students gain interdisciplinary training from participating in regular seminar programs and workshops organized at the Centers.

Box 5: NSF Undergraduate REU at the Four NSF-NIEHS Oceans and Human Health Centers

Beginning in 2004 and continuing for five years, two undergraduates per year work closely with researchers in the NSF-NIEHS OHH Center projects, mentored by the lead project researcher. Students have the opportunity to participate in field sampling and environmental modeling, write reports and publications, and conduct literature searches in specific research areas. They are encouraged to write and present their research at appropri- ate scientific meetings.

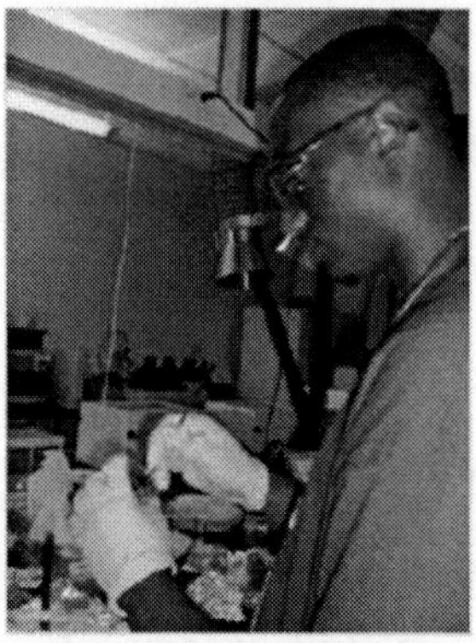

REU students get the opportunity to work directly with OHH researchers in the lab and field. Photo credit: S. Lovelace.

NOAA's Oceans and Human Health Initiative (OHHI)

NOAA is the nation's lead civilian ocean agency; its vision is "an informed society that uses a comprehensive understanding of the role of the ocean, coasts, and atmosphere in the global ecosystem to make the best social and economic decisions." Based on this vision, and a series of reports that called attention to the role of the ocean in affecting human health, Congress appropriated funds to NOAA beginning in 2003 to establish its Oceans and Human Health Initiative (OHHI). OHHI tackles critical issues, such as "are our waters safe to swim in and to drink[3]" and "is our seafood safe to eat," and how we can guard against growing and emerging health impacts while taking advantage of the many benefits the ocean provides.

The OHHI was officially authorized by the OHH Act of 2004 to coordinate and implement NOAA's research and activities related to the role of the oceans in human health and provide support for agency Centers of Excellence in OHH, external research grants, distinguished scholars, traineeships, and a national advisory panel. The OHH Act specified the scientific scope for the OHHI as research related to the agency's mission, including the use of marine organisms as indicators for marine environmental health, ocean pollutants, marine toxins and pathogens, HABs, hypoxia, seafood testing, identification of potential marine products, biology and pathobiology of marine mammals, marine genomics, marine environmental microbiology, ecological chemistry, and conservation medicine. The OHHI was directed to coordinate with the OHH programs of the NSF, NIEHS and other federal agencies and departments.

The mission of the OHHI is to improve understanding and management of the ocean, coasts and Great Lakes to enhance benefits to human health and reduce public health risks. The OHHI investigates relationships between environmental stressors, coastal condi-tion, and human health to maximize health benefits from the ocean, improve the safety of seafood and drink-ing waters, reduce beach closures, and detect emerging health threats. In order to accomplish its mission, the OHHI catalyzes innovative and interdisciplinary OHH research; fosters and facilitates strong partnerships among agencies, academia and the private sector on OHH; and develops and delivers useful tools, tech-nologies and environmental information to public health and natural resource managers, decision-makers and the public.

The goals of the OHHI (see Box 6) are to: (1) lead development of early warning systems to forecast threats and predict long-term risks to human health throughout U.S. coastal and Great Lakes waters; (2) investigate and optimize health benefits from the sea; and (3) help develop a robust oceans and human health community working across disciplines and institutions to improve public health.

OHHI Programmatic Structure

NOAA's OHHI includes competitively awarded internal Centers of Excellence in OHH, external research and internal capacity-development grants, distinguished scholars, and traineeships. Of the funds appropriated to the OHHI, at least 50 percent must be expended in support of the external grants, distinguished scholars, and traineeships. A national advisory panel comprised of top marine and biomedical scientific experts helps guide the development and future direction of the OHHI. The OHHI is founded on partnerships among scientists, public health decision-makers and natural resource managers, and across agencies and academia. All of these program elements work together to accomplish OHHI's mission by coordinating research, outreach, education, and data management within OHHI, across NOAA, and with other agencies to strengthen OHH capacity through partnerships and interdisciplinary research and training.

The OHHI conducts research in nearly every coastal region in the United States, with over 75 OHHI Center- based scientists, 26 external and seven internal grants recipients, and two Distinguished Scholars, working across a multitude of local, state, federal, and academic partnerships. Already, OHHI research activities have resulted in over 80 peer-reviewed scientific publications and have been highlighted at more than 50 scientific conferences, workshops, and other meetings. The OHHI strengthens NOAA's capability to conduct and deliver cutting-edge OHH research and directly supports the "One NOAA" approach and its mission, "To understand and predict changes in the Earth's environment and conserve and manage coastal and marine resources to meet our Nation's economic, social and environmental needs." It also supports three of NOAA's four mission goals: "To protect, restore and manage the use of coastal and ocean resources through ecosystem-based management; to understand climate variability and change to enhance society's ability to plan and respond; and to serve society's needs for weather and water information."

Box 6: NOAA's OHHI Goals and Strategic Objectives

Goal: Early Warning Systems

Lead the development of early warning systems to forecast threats and predict long-term risks to human health throughout U.S. coastal and Great Lakes waters

Strategic Objectives:

- Develop methods, tools, and technologies to predict ocean-related public health risks from pathogens, chemical pollutants, and toxins
- Characterize impacts of coastal ecosystem change, in response to natural factors and environmental stressors such as climate change, land-use and pollution, on ecosystem goods and services related to human health
- Use sentinel species and habitats as integrative indicators of ocean health threats to humans
- Develop and integrate data, tools and technologies applicable to human health into ocean observing systems
- Engage public health and natural resource managers and decision-makers in order to develop and deliver useful products and services
- Evaluate prototype early warning systems and their components for readiness and plan for transition to operations
- Evaluate utility and value of early warning systems and their components to identify, mitigate, and reduce public health risks and related economic and socio-cultural impacts

Goal: Benefits from the Sea

Investigate and optimize health benefits from the sea

Strategic Objectives:

- Assess health benefits and risks of seafood consumption through partnerships with fishery and aquaculture communities
- Partner with subsistence communities to document and evaluate the benefits and risks of marine food products for health promotion and socio-cultural value
- Develop and transfer technology to enhance the healthful characteristics of seafood and minimize ocean-related contamination of seafood
- Increase and leverage partnerships to discover and develop pharmaceuticals, molecular probes, bioactive compounds, and other marine natural products through innovative and ecologically sound exploration and laboratory approaches
- Develop and transfer, to other researchers and the private sector, useful aquatic biomedical models to enhance understanding of human disease processes and provide other health benefits
- Characterize and estimate public health related economic, socio-cultural and broader health benefits (and risks) of coastal and marine resources
- Evaluate utility and value of tools, technologies and information transferred to managers, decision-makers, and other partners

Goal: OHH Community

Develop a robust oceans and human health community working across disciplines and institutions to improve public health

Strategic Objectives:

- Provide opportunities and institutional mechanisms to support interdisciplinary OHH research such as Centers of Excellence, traineeships, early career and distinguished scholar awards, grants and research collaborations
- Partner with the OHH programs of the NSF and NIEHS, as well as other interested agencies, to stimulate scientific interest in OHH across broad institutional and disciplinary boundaries
- Identify key end-users and their needs in order to develop and deliver useful OHH tools, technologies and information
- Communicate findings to decision-makers, managers, scientists and the public
- Provide OHHI information to ocean education programs and media outlets for use in promoting ocean literacy and stewardship
- Partner with other agencies to enable a rapid and coordinated OHH community response to ocean and coastal public health emergencies

NOAA's Centers of Excellence for Oceans and Human Health

Pursuant to the OHH Act of 2004, NOAA designated three national Centers of Excellence in OHH within the agency through an internal competitive process. These Centers, designed to enhance NOAA's internal OHH research capacity, include multi-institutional partnerships with non-NOAA elements and are located in three of NOAA's six line offices – NOAA's Ocean Service, NOAA Research, and NOAA Fisheries. All three Centers are a combination of agency, academic, and other relevant partners, with nearly half of each Center's funding going to its external partners, in addition to the amounts provided to the external community via the OHHI grants, scholars, and traineeships. Combined, the OHHI Centers are working on at least 50 projects through 22 institutional partnerships. In addition to their scientific work, each Center has dedicated outreach personnel and activities, advisory groups, and data management efforts. Also, the directors of the three NOAA OHHI Centers and the four NSF-NIEHS OHH Centers meet at least annually with key staff to share scientific information and research findings, identify opportunities for collaboration, and initiate synthesis and assessment of the implications and application of their research findings.

NOAA Center of Excellence for Oceans and Human Health at the Hollings Marine Laboratory (HML), Charleston, South Carolina

This Center is developing new methods and approaches to identify and characterize chemical and microbial threats to marine ecosystems and human health and to evaluate the health responses of marine organisms to stress (see Box 7). To accomplish these objectives, the Center at HML supports four core research areas: (1) source tracking of marine pathogens; (2) emerging chemical contaminants; (3) applied marine genomics; and (4) monitoring, assessment, and prediction. In addition, scientists at HML are evaluating techniques to determine human health benefits and risks associated with cultured and wild seafood, and to assess the levels and variability of chemical contaminants and beneficial fatty acids in representative finfish and crustaceans.

Pathogen Source Tracking: Scientists are developing novel techniques and sensors for rapidly detecting and tracking marine microbes of public health concern, including bacteria, viruses, protozoan parasites, and HAB species. These tools will be useful for rapidly assessing health risks for humans who recreate in coastal waters or consume shellfish. The technologies allow the sources of microbes to be identified, instead of relying on indicators for contamination, and will provide managers the ability to rapidly respond to beach and shellfish contamination (see Box 26).

Environmental Chemistry: Scientists are conducting research to identify and understand the potentially harmful effects to humans of a variety of chemical contaminants that reach estuaries and the ocean. Cutting-edge laboratory methods and equipment are used to study the fate and effect of compounds of emerging concern, such as new pesticides, herbicides, fire retardants, pharmaceuticals, and personal care products. Development and use of new analytical techniques help identify and quantify the problems, while field, laboratory, and mesocosm experiments allow scientists to look at effects of sublethal concentrations of target contaminants on sentinel marine and coastal species (see Boxes 17 and 21).

Genomic Applications: An interdisciplinary genomics effort allows HML scientists to evaluate physiological responses of organisms to environmental stress from pathogens, toxins, and contaminants at the molecular level. The purpose of the genomics program is to provide insights into the physiological status of organisms and to translate this into an understanding of overall population health. Development of this technology for key marine species, including the bioinformatic tools to interpret the complex data, will provide an additional tool for monitoring and assessing ecosystem health (see Box 32).

Monitoring, Assessment, and Prediction: Scientists are evaluating the performance of environmental quality and public health indicators developed by other OHHI program elements using southeastern tidal creeks as sentinel habitats. Findings are used to develop approaches for simultaneously monitoring and assessing the ecological condition and public health status of coastal environments (see Box 29).

HML is operated as a partnership comprised of NOAA, the National Institute of Standards and Technology (NIST), the South Carolina Department of Natural Resources, the College of Charleston, and the Medical University of South Carolina. This unique arrangement integrates basic, applied, and biomedical scientists into multidisciplinary research teams and establishes programs that link environmental conditions in the coastal zone to human health and socio-economic well being.

BOX 7: DEVELOPMENT OF TOOLS TO RAPIDLY DETECT AND IDENTIFY HARMFUL MARINE ORGANISMS

Scientists at HML have developed cutting-edge technologies to rapidly detect and determine the source of waterborne pathogens such as *E. coli*, viruses, and protozoa that can cause disease in humans. Some of these harmful microorganisms are indigenous in coastal waters, while others are introduced through agricultural and wastewater inputs, storm water run-off, and the excrement of domestic and wild animals. Current technologies are slow and rely on indicators of the harm-ful marine organisms. These new methods can rapidly assess the actual presence of pathogens in coastal oceans and pin down their source. Also, in response to the need of public health and environmental managers to quickly screen for the presence of harmful algal species, HML researchers have developed a new technology to simultaneously detect multiple algal species. Known as SIVCA (Species Identification via Chimeric Amplification), this technology is a vast improvement over existing methods. It requires less time and expertise to operate, and is cheaper. In just a year and half, these tools have gone from concept to field trials with hopes of having the technology broadly available to local public health professionals within the next several years. With these advanced DNA technologies in hand, managers will be equipped to determine when beaches are safe for swimming and seafood safe to eat, and to mitigate pollution at its source.

OHHI researchers are developing new DNA-based technologies for public health and resource managers to rapidly detect and determine the source of harmful algal species and waterborne pathogens such as *E. coli,* viruses, and protozoa. Photo credit: S. Lovelace.

In addition, HML maintains a Center-wide data manage ment program to develop the information technology, infrastructure, and expertise required to share, synthesize, integrate, and archive the data and information produced by the Center, its partners, and other collaborators. The education and outreach program communicates and transfers the Center's knowledge and technology to natural resource and public health managers, scientists, regulatory agencies, local decision-makers, university students, teachers, and the public.

NOAA Center of Excellence for Great Lakes and Human Health (CEGLHH) at the Great Lakes Environmental Research Laboratory, Ann Arbor, Michigan

This Center focuses on understanding the inter-relationships among the Great Lakes ecosystem, water quality, and human health. The Center employs a multidisciplinary approach to understand and forecast coastal- related, human health impacts for natural resource and public policy decision-making, and develop tools to reduce human health risks associated with three research priority areas: beach closures, HABs, and drinking water quality.

Many processes and events influence the sources, transport, and loading of pollutants, bacteria, and nutrients to the Great Lakes, such as land use and weather. Defining and forecasting these relationships are the primary goals of the Center. Research is concentrated on providing predictions of coastal water quality that can be used directly to reduce risks to human health associated with recreational exposure and human consumption of Great Lakes water. Main research foci include: (1) bacterial and microbial pollutant sources and loadings; (2) pathogen/virus transport (see Box 14); (3) microbial research; (4) nearshore transport; and (5) HABs (see Box 18). The research includes laboratory work, field experimentation, and computer modeling.

The Center is housed within and led by the NOAA Great Lakes Environmental Research Laboratory (GLERL), but is comprised of multiple federal agencies, universities, and public health networks. GLERL's partners include Michigan State University, EPA Chicago, USGS, EPA Athens, GA, Florida Institute of *Oceanography*, NOAA Beaufort Laboratory, NOAA NOS Silver Spring, Michigan Sea Grant Extension, University of Michigan, and the Great Lakes Human Health Network.

CEGLHH uses a multidisciplinary approach to translate scientific information and research into materials to aid public health officials, beach and coastal zone managers, drinking water operators, local and tribal governments, and communities in making sound environmental decisions (see Box 8). In addition, CEGLHH works to assess user needs by collaborating with stakeholder groups (see Box 11). Outreach activities are coordinated through the well-established Great Lakes Sea Grant Extension Program and include a broad range of activities, such as needs assessment workshops, a public access Web site, newsletters, information-sharing partnerships, publications, public presentations and training workshops.

BOX 8: NEW BEACH FORECAST MODEL FOR THE GREAT LAKES

NOAA's CEGLHH and the USGS teamed up to implement a forecast model that can predict recreational beach conditions in the Great Lakes to enable public health officials and beach managers to better manage beaches and reduce health risks. Project SAFE (Swim Advisory Forecast Estimate) is a model that predicts bacterial counts for *E . coli* based on rainfall, wave heights, and the direction of lake currents, to determine when counts are high enough to threaten human health. This model is one of the first recreational water health forecasting system in the country and does not rely on time-consuming sampling that tells beach managers today that they should have closed the beaches yesterday. CEGLHH provided numerous inputs to help develop and sustain the model, ranging from technical support to hydrological data and financial support. The lab also provides the necessary daily forecast on lake currents and wave heights. This model became operational during summer 2006 in Indiana's Lake Michigan beaches and effectively predicted *E. coli* concentrations. Predicted bacterial counts were posted through hypertexts in local newspapers, the USGS and Project SAFE Web sites, and were sent to beach managers, Gary Sanitary District, and Indiana Department of Environmental Management by 10 AM each day, which gave beach managers enough time to inform the public of beach advisories or closures. Project SAFE, in conjunction with the onset of rapid sampling techniques, will greatly improve the prediction and detection of harmful microorganisms such as *E. coli* in the Great Lakes and could serve as a model for application in other coastal regions of the United States.

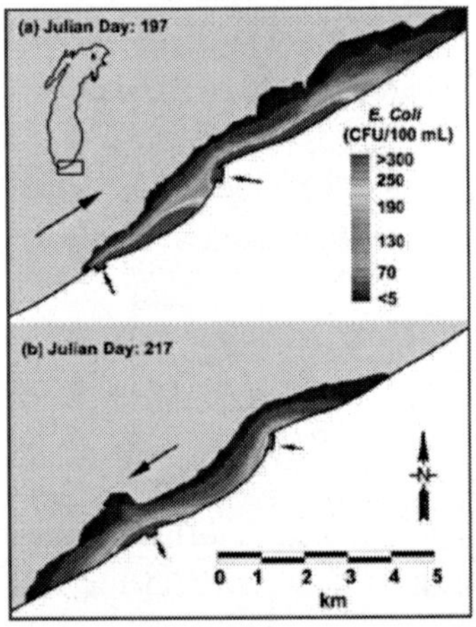

Dangerous levels of bacteria, such as *E. coli* (shown here in red and orange), can be predicted using one of the first operational forecast models to allow for daily updates to Great Lakes beach managers. Photo credit: P. Mantha and J. Rose.

NOAA West Coast Center of Excellence for Oceans and Human Health (WCCOHH) at the Northwest Fisheries Science Center, Seattle, Washington

This Center has strong research programs with proven track records in a wide range of scientific fields (e.g., climatology, *Oceanography*, microbiology, genetics and molecular biology, immunology, ecotoxicology, neurotoxicology, developmental biology, plankton ecology, physiology, and marine mammal ecology). The WCCOHH conducts its research through four core programs: (1) pathogens, viruses, and bacteria; (2) chemical contaminants and biotoxins (see Box 19); (3) marine mammals and fish as sentinel organisms (see Box 31); and (4) climate impacts. Key priorities for the Center include sharing data and research results with other institutions and the public, fostering the exchange of information among diverse communities, including other OHH programs, and providing educational opportunities.

BOX 9: RISK ASSESSMENT TOOLS TO DETECT PATHOGENIC *VIBRIOS* IN SEAFOOD

Scientists at the WCCOHH at the Northwest Fisheries Science Center have developed a rapid DNA test to measure pathogenic strains of *Vibrio vulnificus*, which are capable of causing fatal infections in people who consume raw shellfish. Although molluscan shellfish, such as oysters, can harbor hundreds of individual strains of this naturally occurring marine bacterium, it has long been recognized that only a low percentage of strains are capable of causing life-threatening infections in people. This has been a major obstacle to being able to rapidly determine the actual safety of shellfish for human consumption. Until the identification of the genetic marker targeted by this DNA-based method, there has been no tool available to differentiate the pathogenic *V. vulnificus* isolates from the majority of non-pathogenic environmental strains. The USDA and FDA have recently used the method to survey market oysters to assess the relative percentage of pathogenic *V. vulnificus* strains in shellfish. Furthermore, this technology has led to a university-federal government collaboration to carry out the genome sequencing of several *V. vulnificus* strains, which will then be used to identify the genes or other factors that cause some *V. vulnificus* strains to be more pathogenic than others. Such advanced predictive technologies will enable public health managers to better determine the safety of seafood and not indiscriminately close shellfish beds based only on identifying *V. vulnificus* without differentiation between being pathogenic or non-pathogenic strains.

Using new DNA tools, pathogenic *Vibrios* can be distinguished from non-harmful strains, enabling better management of shellfish beds and securing the safety of shellfish such as oysters. Photo credit: M. Strom.

The WCCOHH's research will (1) provide a better understanding of how pathogens and toxins, both natural and human-made, interact with and affect shellfish and fish, enhance the development of tools to identify direct risks that contaminated shellfish pose to consumers, and develop methods to forecast and mitigate threats to human health. Conversely, indirect effects to human health can be realized by understanding the health of sentinel species, like fish and marine mammals. In addition WCCOHH's research will provide a better understanding of the transmission dynamics of infectious diseases and their interactions with other stressors, such as toxins and environmental variability (see Box 9).

Key partners of the WCCOHH include the University of Washington (PNW H2O Seattle, WA), Institute for Systems Biology (Seattle, WA), Oregon State University (Corvallis, OR), the Marine Mammal Center (Sausalito, CA), the University of California, Davis (Davis, CA), and NOAA's Alaska Fisheries Science Center (Seattle, WA). Also, the WCCOHH has external collaborations with researchers affiliated with the University of Florida and the National Research Council, Canada.

WCCOHH educational and outreach activities emphasize OHH products and services, strengthen awareness of our connection to the oceans, and increase environmental stewardship. Specific activities include participation in major scientific conferences, educator workshops, annual investigator meetings, and local community events; development of a newsletter and interactive Web site; and sponsorship of an educator at sea and student travel to a national conference.

OHHI External Research Grants

The OHHI External Grants Program engages the non-federal research community in investigations that directly complement NOAA's ongoing OHHI activities. NOAA is the only agency that has a targeted external research grant opportunity in the OHH arena. In keeping with the OHH Act requirements, only non-federal applicants are eligible to apply for funds allocated to this program element. The grants are structured to provide useful information for public health and natural resource managers related to NOAA's OHHI mission (including Centers). Grantees must demonstrate the ability to perform and sustain interdisciplinary and applied research and execute outreach and data management plans.

Through 2006, the OHHI External Grants Program has received over 260 proposals in response to its Requests for Applications and has granted 26 competitive awards to investigators at academic institutions throughout the United States: Northeast (9), Southeast (includes the Gulf of Mexico) (7), Pacific Northwest (2), Southwest (3), Great Lakes (3), and Pacific Islands (2). Topically, work supported by the grants covers the full OHHI scientific scope, including research on ocean-related pathogens (see Boxes 10 and 15), toxins, chemical contaminants, benefits and risks of seafood consumption, sentinel species, marine natural products, and the effects of multiple ecosystem stressors on OHH. Regular Principal Investigator meetings provide opportunities for external grantees and NOAA scientists to compare findings, share information and work plans, and develop new collaborations.

OHHI Internal Capacity Development

In addition to external grants, the NOAA OHHI has supported some small competitive awards within NOAA through its Internal Capacity Development Program. This program element is designed to stimulate OHH activities across NOAA, catalyze cross-NOAA collaboration on OHH issues, capitalize on existing strengths, and take advantage of the OHH-related scientific capacity residing in NOAA entities outside the Centers. Focus areas for this peer-reviewed competition resulted from an internal NOAA OHH workshop in May 2004, which drew together scientists working on oceans and health issues throughout the agency. In contrast to the external grant program, these funds are directed toward internal NOAA research projects, though outside collaboration is possible. In 2005, 21 proposals were received, seven were awarded. These awards have focused primarily on sensor development, chemical contaminants, sentinel species, and bioinformatics.

BOX 10: ZOONOTIC DISEASE SURVEILLANCE: USING SEABIRDS AND MARINE MAMMALS TO DETECT HEALTH THREATS

An OHHI external grant recipient at the Woods Hole Oceanographic Institution, along with colleagues and partners at Cornell University, Tufts Cummings School of Veterinary Medicine, and the Cape Cod Stranding Network, are addressing worldwide concerns about animal vectors of human disease (zoonoses) by learning more about the role of marine mammals and seabirds as reservoirs for disease. This team investigated the prevalence of zoonotic bacteria, parasites, and viruses in the southwestern Gulf of Maine in whales, dolphins, seals, gulls, eider ducks, and loons and screened a total of 116 normal and diseased individuals from live captures, fishery bycatch, or beach strandings. Representatives of 79 of approximately 200 potentially zoonotic pathogens worldwide were detected in the marine organisms. Infectious agents associated with diarrhea and other chronic diseases such as brucellosis (which can cause fever, sweats, headaches, back and joint pains, physical weakness and infections of the central nervous systems or lining of the heart) were detected in 26 percent of the cases. Other particularly noteworthy findings included detection of a Brucella-like agent in birds for the first time, a broad array of pathogens in a deep-diving beaked whale, and marked antibiotic resistance patterns in bacteria from live, bycaught, and stranded marine mammals and birds.

Harbor seals and sea birds can carry human diseases and provide clues about how the spread of current and emerging threats in the oceans can affect human health. Photo credit: M. Moore.

OHHI Distinguished Scholars

The Distinguished Scholars Program builds NOAA's capacity by bringing world-renowned scientists to work with NOAA on cutting-edge ocean and health sciences and their application, and by recognizing forward-looking scientific leadership in the emerging interdisciplinary OHH scientific field. Proposals are reviewed based on scientific merit, collaboration with NOAA, and ability to add significantly to NOAA's OHH research capabilities. Based on a peer-reviewed competition, the OHHI awarded its first two

Distinguished Scholars grants in 2006 to Dr. Rita Colwell (University of Maryland at College Park and Johns Hopkins University Bloomberg School of Public Health and former Director of the National Science Foundation) and Dr. Phillip Roberts (Georgia Institute of Technology). Dr. Colwell is focusing on application of genomic technologies for disease surveillance in the marine environment and the development of integrated, global observations for disease surveillance. She is working with the OHH Center at the HML and with the OHHI Program Management Team. Dr. Roberts is focusing on improving models for pathogen forecasting in nearshore environments in the Great Lakes. He is working principally with the OHH Center at GLERL.

OHHI Traineeships

NOAA's OHHI Traineeships are designed to build a new cadre of scientists working in the OHH field that are trained across disciplines and institutions to improve public health. These awards provide cross-disciplinary training opportunities, build institutional support for OHH research and application, enhance communication with public health and natural resource managers, stimulate scientific advances in OHH, and acquaint young researchers with NOAA OHH research. OHHI trainee- ships will be awarded on a competitive basis for pre-doctoral students and post-doctoral researchers who will become actively involved in OHH research.

OHHI Public Information and Outreach

The OHHI Public Information and Outreach Program strives to engage and involve public health and natural resource managers and decision-makers, and the broader OHH scientific community in the application of OHHI research. This program helps to disseminate and communicate OHHI research findings, assessments, tools, and technologies to managers and decision-makers, other users, and the public (see Box 11). Public information and outreach activities are conducted at both the national level and through the OHHI Centers. Each Center has a coordinator who develops partnerships, leverages resources, and organizes and implements targeted activities. Also, all OHHI grant recipients are required to have an outreach component. Program activities are conducted in consultation with other federal agencies and in cooperation with regional Sea Grant College Programs, as well as with appropriate state and local partners. Activities range from workshops, training opportunities, internships, seminars, scientific meetings, and other public events, to the development of Web sites, informational products, and curricula. This program also helps to enhance public literacy and understanding about the effects of the oceans on human health, promote ocean stewardship, and build a new cadre of interdisciplinary OHH professionals. (See Box 41).

BOX 11: NEEDS ASSESSMENT WORKSHOP TO HELP DIRECT BEACH HEALTH RESEARCH

In November 2005, the CEGLHH partnered with the USGS, EPA, and Great Lakes Beach Association to host a workshop to identify needs and concerns of beach managers, public health officials, and other stakeholders. The workshop's purpose was to define research priorities for addressing recreational water quality issues in the Great Lakes related to beach closures due to pathogen pollution or HABs. All Great Lakes states were represented among the over 40 participants. Workshop outcomes included establishment of a steering committee to ensure implementation of assessed needs, a beach management training course, a training video on the use of models to predict bacteria levels, and a standardized sanitary survey tool to identify contamination sources at beaches.

Needs assessment workshops help identify OHH tools and information beach and seafood managers can use to reduce human health threats. Photo credit: S. Joseph.

OHHI Data Management

To ensure maximum consistency, access, and use of data derived with OHHI funding, the OHHI established a Data Management Working Group, which includes active participation of experts from each of the OHHI Centers and the OHHI Program Office. This group is design-ing a seamless, distributed OHHI data/metadata infrastructure to provide access to data and related information products by OHHI researchers, the general scientific community, and operational user communities (e.g., public health and natural resource managers, homeland security). Specific data management objectives include developing and implementing policies and protocols throughout the OHHI and in coordination with other OHH programs, such as the NSF-NIEHS OHH Centers, to maximize access and minimize duplication. Also, the Data Management Working Group will coordinate with other NOAA data management activities such as IOOS, the Data Management Integration Team, and other ocean, coastal, and public health surveillance systems to develop long-term useful data streams for monitoring and analysis. The goal of the OHHI data management policy, in principle, is to allow free and open access to the research community but still provide scientists a restricted initial access period for the analysis, research, and publication of their data. Proper acknowledgement and citation of all data used will be required in publications. OHHI is making arrangements to ensure that "orphan" data sets (i.e., smaller regional networks and local observations that may not be routinely saved or readily available) will be archived.

OHHI Partnerships

Partnerships at the federal, state, local, and eventually international levels are critical elements of the OHHI. To ensure synergy and avoid unnecessary duplication, the OHHI works very closely with several other NOAA programs and with numerous other federal agencies, including all of those represented on the IWG-4H. Most state and local partnerships are developed at the regional level through the OHHI Centers of Excellence and/or by the external grant investigators.

Understanding the connections between marine animal health and human health, and identifying key species as "sentinels" of potential human health impact, requires a close working relationship with NOAA's Marine Mammal Health and Stranding Response Program (MMHSRP). This Program focuses on marine protected species and marine animal stranding. The collaboration includes joint research, planning, and funding, as well as post doctoral training.

The OHHI works closely with the ECOHAB (Ecology of Harmful Algal Blooms) Program (supported jointly by NOAA, NSF, EPA, NASA and ONR) and NOAA's MERHAB (Marine Emergency Response to Harmful Algal Bloom) Program, with a principal focus on work related to the human health impact of HAB-related bio-toxins. Within NOAA, the programs coordinate research planning and funding of external research grants.

In addition to the OHHI, NOAA has several programs that have at least some focus on science and technology related to marine natural products and pharmaceuticals. The Ocean Exploration (OE) Program will be taking samples of opportunity during ship runs, and may also offer competitive research grants. One of the National Undersea Research Program (NURP) Centers has expertise in marine pharmaceuticals research, development, and sample archiving. In previous years, the National Sea Grant College Program has provided grants to support research on natural products from the sea, including pharmaceuticals. In addition to its role as an OHHI Center, HML has capabilities in marine natural product analysis. The OHHI is leading an internal NOAA effort to coordinate approaches to the discovery and identification of new marine natural products and pharmaceuticals. The NIH National Cancer Institute is also part of this planning process.

The OHHI collaborates and coordinates activities with the NSF-NIEHS, including annual meetings of the directors of all seven OHH Centers. The OHHI also works with EPA at the national and regional levels on beach and water quality management and with the USGS on regional issues related to beach management and risk forecasting and at the national level on marine animal health and surveillance issues. The OHHI connects with CDC through joint research planning and the transfer of research results to the public health community.

Although the OHHI's main focus is on the United States, it has supported a bit of international work related to malaria in Africa. Collaborations with other international health and Earth science organizations could be strengthened in future years.

Related OHH Programs in Other Agencies

The **Centers for Disease Control and Prevention (CDC)** is one of the 13 major operating components of the Department of Health and Human Services (HHS). Since it was founded in 1946 to help control mala ria, CDC has remained at the forefront of public health efforts to prevent and control infectious and chronic diseases, injuries, workplace hazards, disabilities, and environmental-health threats. CDC applies research findings to improve people's daily lives and responds to health emergencies (see Appendix 2: Agency Prospectus).

CDC is conducting studies to collect data that will improve public health decision-making and responses to ocean-borne health threats. Since 1998, the Health Studies Branch at the National Center for Environmental Health (NCEH) has been funded to support a cooperative agreement with six eastern seaboard states relative to the harmful alga, *Pfiesteria piscicida,* and more recently other HABs. The program includes biological research, environmental exposure assessments, human disease surveillance, clinical evaluations of ill people, environmental monitoring, and the development of state HAB response plans. NCEH participates in collaborative HAB-related research activities with other federal agencies, universities, and state and local public health agencies and developed the HAB-related Illness Surveillance System (HABISS) (see Box 25). An ongoing epidemiological study involving a partnership between CDC and NIEHS demonstrated that people exposed to aerosolized HAB toxin (specifically brevetoxin associated with Florida red tides) experience respiratory problems and that these problems are of special concern for people with asthma. The CDC team developed toxin assays and other methods and produced a targeted public health protection program based on its findings. CDC is also leading partnership studies investigating human exposures to freshwater HAB toxins in the Great Lakes region.

In related work, CDC is building a National Environmental Public Health Tracking Network that will provide scientific information on environmental exposures and adverse health conditions as well as the possible spatial and temporal relations between them. This network will include exposures and health effects of marine origin. Other OHH-related activities include service on the National HAB Committee and the National Advisory Panel for NOAA's OHHI, both of which help focus OHH research efforts.

CDC's National Center for Infectious Diseases (NCID) conducts various surveillance programs for food- and water-borne diseases reported by state health agencies and in cooperation with the USDA and the FDA. This surveillance includes illnesses caused by *Vibrio* bacteria (most of which are attributable to consumption of contaminated seafood) and marine toxins, which account for about four percent of the food-borne disease reports. NCID is working with NCEH to improve waterborne disease surveillance and also collaborates with the EPA to assess public health risks from water-related recreational activities, which include recreational exposures to toxins, contaminants and pathogens.

The CDC's National Institute for Occupational Safety and Health is responsible for conducting research and making recommendations for the prevention of work- related injuries and illnesses, including those associated with commercial seafood harvesting.

The **U.S. Environmental Protection Agency (EPA)** is an independent federal agency whose mission is to protect human health and the natural environment. EPA develops and implements federal regulations to protect human health and issues non-regulatory health advisories to warn the public about levels of contaminants in the air, land, and water, including marine waters, that can cause harm.

The EPA Office of Research and Development (ORD) has initiated a multi-year planning effort for priority research. Several of the EPA multi-year plans (MYPs) pertain directly to OHH or OHH-related topics, including drinking water, water quality, human health, human health risk assessment, and global change. All MYPs can be accessed online at http://epa.gov/osp/myp.htm. Research efforts of particular relevance to OHH include studies on how water-borne contaminants may affect swimmers and beachgoers. For example, ORD conducted epidemiological studies on three freshwater (Great Lakes) beaches to evaluate molecular methods for determining when beaches should be closed due to pathogen contamination. Related studies were initiated in the Gulf of Mexico in 2005, but these were prematurely terminated by Hurricane Katrina impacts. A new EPA supported OHH-related research program, being developed through the agency's National Center for Environmental Research, will evaluate relationships between changes in biodiversity and human health.

Domestically, EPA participates in numerous interagency partnerships with OHH-related themes, including: with NOAA, ONR, and NASA in the (ECOHAB) Program; with NIEHS' National Toxicology Program to evalu-ate toxicity of contaminants in drinking water; with CDC on surveillance of water-borne diseases; with the U.S. Army Corps of Engineers for wetland conservation and restoration; and with NOAA and NSF to evaluate Hurricane Katrina's impacts on Gulf of Mexico coastal waters. EPA representatives serve on the National HAB Committee and on the National Advisory Panel for NOAA's OHHI.

EPA's international programs help protect the environment along our common borders with Mexico and Canada, reduce global environmental threats such as marine pollution and toxic chemicals, integrate environmental protection with international trade and investment, and exchange innovative practices and technologies. These activities not only help sustain the health of the global ocean, but also improve environmental quality across the nation. In addition, EPA provides U.S. leadership related to international marine pollution issues.

The **Food and Drug Administration (FDA)** of the Department of Health and Human Services ensures the safety of food and drugs available to consumers, and is the agency primarily responsible for approval of pharmaceuticals. While essentially a regulatory agency, the FDA conducts and encourages research to support its mission.

The FDA has several OHH-related research programs, many at the Center for Food Safety and Applied Nutrition (CFSAN), pursuant to its responsibility for the safety of shellfish and other seafood. CFSAN's goals are to understand the risks and develop cost-effective strategies for addressing those risks and practical tools for implementing those strategies, so that the FDA can minimize risks and maximize benefits of seafood con-sumption. One CFSAN concern is improper handling of seafood products, which can allow them to decompose. Some forms of decomposition can lead to the formation of toxins that cause human illness and, in general, decomposition lowers the value of increasingly scarce marine resources. CFSAN conducts research to establish time/temperature parameters for harvest, transportation, and storage of seafood that will ensure its wholesomeness, and develops methods for detecting decomposed seafood. In addition, absorption of the volatile mono- and bicyclic aromatic hydrocarbons from petrochemical contamination leads to "petrochemical taint." CFSAN research on petrochemical contamination will introduce and advance chemical analytical methods for the detection and quantification of petrochemical taint in seafood.

In the mid-1990s, the FDA implemented a Hazard Analysis Critical Control Point (HACCP) program for the seafood industry to analyze hazards, identify and monitor critical control points, establish preventive measures, establish corrective actions to be taken when monitoring shows that a critical limit has not been met, and verify that the system is working and documented properly. The FDA published the Fish and Fishery Products Hazards and Controls Guide to assist processors in the development of HACCP plans; this guide is updated periodically to reflect advances in science-based knowledge regarding seafood hazards. The safety features of FDA's seafood HACCP regulations have also been incorporated into NOAA's National Seafood Inspection Program.

Research on marine toxins includes development of toxin-detection methods, characterization of toxins and emerging vectors of toxins, and exploration of better strategies for managing the risk of natural toxins in seafood. The latter includes the Signal Environmental And Plankton Observations in Real Time (SEAPORT) program (see Box 42) and participation in preparation of major HAB reports.

Aquaculture is a rapidly growing industry worldwide, contributing significantly to the global food supply. Because of industry expansion and trends toward high- intensity aquaculture, diseases frequently occur in farm- raised fish or shellfish. Many of these diseases can be addressed with drug therapy. Therapeutic drugs and chemicals are available to the global industry for treating disease, but their use raises several concerns, including the development of resistant strains of microorganisms, environmental impacts of drug residues, and drug residues in products. Regulations governing aquaculture drug use vary from country to country, so the detection of residues in a product is important for ensuring compliance. The FDA has identified a priority list of aquaculture drugs and chemicals for analytical-method development in specific applications based on food safety and the extent of known use within the global aquaculture industry.

Pathogenic microorganisms present in the aquatic environment can be accumulated by shellfish, posing a serious risk to consumers of raw or inadequately cooked shellfish. The majority of shellfish-associated illnesses are attributed to enteric viral pathogens. Research on pathogenic viruses and bacteria is conducted by the Office of Seafood (OS) and Office of Applied Research and Safety Assessment (OARSA). Current OS research activities focus on: (1) the ecology of indicator microorganisms of fecal pollution in estuarine waters and their interaction with shellfish and (2) the development of methods for the extraction and detection of enteric viruses in shellfish tissue. Findings will be evaluated to determine if shellfish growing-area classifications are suitable for indicating the presence of enteric viruses and to determine if any of the indicator organisms could be used to predict the risk of viral contamination. Research conducted by OARSA seeks to identify and characterize virulence determinants/genes of food-borne pathogens, such as marine *Vibrio* species. This work will help lead to development of gene-based tests that are specific for pathogenic organisms, thus allowing regulatory action to be narrowly focused on actual risk, the removal of contaminated food from the marketplace and steps to prevent future contamination.

In its role in approving pharmaceuticals, the FDA's Center for Drug Evaluation and Research (CDER) has approved some marine-derived drugs.

The **Marine Mammal Commission (MMC)** is charged with developing, reviewing, and making recommendations on domestic and international actions and policies of all federal agencies with respect to marine mammal protection and conservation, including the health and stability of marine ecosystems upon which the mammals depend (i.e., ocean health) and the health of native subsistence harvesters and consumers of marine mammals (i.e., human health).

With respect to the OHH-related activities, the MMC is particularly interested in: (1) the relationship between measures of ocean health and marine mammal biology, ecology, population dynamics, and unusual mortality events; (2) the efficacy of using marine mammals as sentinels of ocean and human health; (3) the potential communication of diseases from humans and domestic or feral animals to marine mammals; (4) the potential for marine mammals to communicate diseases to subsis-tence hunters and others who may come into contact with marine mammals; and (5) the relationship among ocean health, contaminant loads in marine mammal tissues, and the health of subsistence hunters.

Although the MMC has supported some research projects relevant to the five issues listed above, it does not have the resources to establish a research program focused solely on OHH issues. Historically, MMC supported some of the earliest work on responses to marine mammal stranding events and participated in the development of a specimen bank of marine mammal tissue samples that can be used in various health-related studies. Currently, MMC participates in a NOAA Fisheries working group on marine mammal unusual mortality events. In addition, the MMC sponsored both a workshop dealing with prevention of potential viral epidemics in Hawaiian monk seals and an assessment of the risk of illness and injury among marine mammal workers and the public who may come into contact with stranded animals. Fifty percent (243/483) of respondents reported suffering an injury caused by a marine mammal; 23 percent reported having a skin rash or reaction. Common marine mammal, work-related illnesses included conjunctivitis, several types of dermatitis, and "seal finger" (apparently a mycoplasma infection). Severe illnesses were less common, but included tuberculosis, leptospirosis, and brucellosis.

The **National Aeronautics and Space Administration (NASA)** is an independent federal agency whose mission is to pioneer the future in space exploration, scientific discovery, and aeronautics research. NASA uses observations and modeling to help answer scientific questions about the ocean and its role in the Earth system.

NASA's OHH activities include basic Earth science research and the extension of NASA Earth science research results to operational partner agencies' public health and coastal management decision-support tools in areas including toxic HABs. NASA supports a cooperative agreement between NRL and Applied Coherent Technologies, Inc. to support NOAA HAB activi-ties through the Research, Education and Applications Solution Network (REASoN) project, which integrates measurements from NASA and NOAA satellites, available coastal observations, and coastal ocean model outputs into the NOAA HAB Bulletin and NOAA Harmful Algal Blooms Observing System (HABSOS). NASA's Ocean Biology and Biogeochemistry research program, through their ECOHAB partnership with other federal agencies, funds basic HAB research on optical propreties of *Karenia brevis* and its detection in the Gulf of Mexico, mycosporine-type amino acids as markers for harmful dinoflagellates, and the coupled biological and physical dynamics of HABs in waters off of the Pacific Northwest. NASA's Applied Sciences Public Health Program supports projects to enhance partner agencies' decision-support tools with NASA Earth science research results for societal benefit. The Applied Sciences Public Health Program will be expanding its OHH activities by participating in the NASA Research Opportunities in Space and Earth Sciences (ROSES) solicitation in FY07 and encouraging proposals related to OHH. In the past, NASA has also partnered with NOAA, EPA, NSF, and EPRI (formerly the Electric Power Research Institute) in the multi-agency Climate Variability and Human Health program.

The **U.S. Geological Survey (USGS),** a bureau of the U.S. Department of the Interior, provides reliable scientific information to describe and understand Earth; minimize loss of life and property from natural disasters; manage water, biological, energy, and mineral resources; and enhance and protect our quality of life.

The USGS conducts a diverse range of programs and activities that provide OHH-related research and information products. Many of these investigations are leveraged by support from partnering among federal and local government agencies, and they address OHH-related issues such as impacts of storms, tsunamis, earthquakes, landslides, climate change, and sea-level rise on coastal communities, as well effects of contaminants, biotoxins, and pathogens on humans.

USGS programs monitor the quality and quantity of water in the nation's rivers, including water and constituent delivery to estuarine and coastal ecosystems; evaluate the sources and mechanisms of contaminant production and delivery; and develop methods to predict and manage beach conditions. In marine, coastal, and Great Lakes environments, the USGS also investigates the condition of wetland, estuarine, coral, and offshore marine habitats and the health of shorebirds and water birds, sea ducks, pelagic seabirds, marine mammals, sea turtles, and fish stocks. These studies provide a means to observe early warning of issues that may affect human health, such as loss of the role of ecosystems in maintaining a healthy coastal environment, depleted food sources, or emerging environmental contaminants, which are of specific concern to the OHH programs. The USGS also uses remote sensing to monitor coastal land-use changes, studies the connection between people and those changes, and provides information that can be used to address the consequences of those changes, including information related to human health. In conducting these studies, the USGS works with a wide range of federal, state, and local agencies.

OHH International Partnerships

Many of the individual agencies collaborate, to varying degrees, with international organizations and countries on work related to the OHH topics covered by this interagency program. While these relationships have evolved over time, and some are well established, an important next step will be to further develop and expand existing relationships and improve flow of environmental data and health information worldwide. Key international organizations with whom the Interagency OHH Program may interact include: International Atomic Energy Agency (IAEA); World Organization for Animal Health (OIE); World Health Organization (WHO); Food and Agriculture Organization of the United Nations (FAO); International Whaling Commission (IWC); Global Earth Observation System of Systems (GEOSS) including the Group on Earth Observations (GEO); United Nations Educational, Scientific and Cultural Organization (UNESCO), including International Council for the Exploration of the Sea (ICES); Intergovernmental Panel on Harmful Algal Blooms of the International Oceanographic Commission (IOC); North Pacific Marine Science Organization (PICES); Arctic Council and its Arctic Marine Assessment Program (AMAP); the

International Geosphere-Biosphere Programme (IGBP); and countries with whom the United States has multi- bilateral environ mental and research agreements..

3: Opportunities for Major Advances in OHH Research and Application

This chapter is the first comprehensive effort to address ongoing OHH research work and priorities at national and interagency levels. It describes the breadth of priority on-going activities that will advance OHH research and application in the nine agencies engaged in the Interagency OHH Program. The lists presented here were developed by the IWG-4H based in part on input received via several pathways, including the JSOST-sponsored Ocean Research Priorities Plan Public Workshop in Denver, CO (April 2006), special meetings of the NOAA and NSF-NIEHS OHH Center Directors and other representatives in Honolulu, HI (February 2006) and Seattle, WA (April 2006), and a variety of written materials provided by numerous experts in their respective OHH fields. The opportunities for advancement are organized under three sub-headings: (1) Research Priorities; (2) Infrastructure; and (3) Transition of Research to Application, and include highlights of selected agency activities (see Box 12). Further prioritization is provided in Chapter 4.

Priority Research Areas

Pathogens, Potential Pathogens, and Indicators

Ocean and estuarine ecosystems can directly and indirectly impact the extent to which humans are exposed to disease-causing organisms known as pathogens, including bacteria, viruses, protozoa, and fungi. One of the most significant ocean health threats to humans is exposure to recreational waters and drinking water contaminated with pathogens (see Boxes 13, 14). Sources of pathogens in ocean waters may include run-off from agricultural and animal production systems into estuaries, rivers, and creeks; wastewater inputs associated with sewage discharge, leaking septic systems, and stormwater run-off from urban and suburban areas; and discharges of ship ballast waters. Currently, the EPA, CDC, and Council of State and Territorial Epidemiologists track food-borne diseases and outbreaks. However, the system is passive and compliance with the request to report outbreaks is incomplete. Thus, the number of food- and water-borne disease outbreaks is likely under-report-ed. In addition, because many of the illnesses associated with food and water contamination have symptoms in common with other illnesses, such as the flu, they are often misdiagnosed.

Pathogen research in coastal environments has advanced considerably under the Interagency OHH Program, and in response to the Beaches Act of 2000, which directed the EPA to develop new technologies to rapidly measure water quality and improve understanding of the relationship between water quality and health of swimmers. Other advances from work in the NOAA and NSF-NIEHS OHH programs and in several other agencies have resulted in expanded capabilities to detect, identify, determine the source of, and eventually forecast the occurrence of pathogens in coastal waters, shellfish, and drinking waters (see Boxes 15, 16). These advances in pathogen research will likely lead to development of new biological sensors that can be incorporated into ocean observing systems. Also, advancements in DNA technologies have allowed for discovery of genetic factors associated with virulence of *Vibrio* and other bacteria, so that harmful strains can be identified and researchers can differentiate human from nonhuman sources of contamination. When fully developed, new tools based on these advances will provide faster and more reliable information to protect consumers from contaminated seafood (see Seafood Safety Section).

Box 12: Primary Areas for Advancement in the Interagency OHH Research Program

Research Priorities

Research Areas:

- Pathogens
- Chemical contaminants
- Harmful algal blooms
- Seafood safety
- Pharmaceuticals and other beneficial products

Cross-Cutting Areas:

- Epidemiology
- Sentinel species/habitats and biological models
- Genomics and other "omics" technologies
- Social, behavioral and economic studies
- Climate change

Infrastructure

- Linking to ocean observing systems
- Data management and access
- Standards and standardized methods
- Access to the sea: platforms for OHH sampling and exploration

Transition of Research to Application

- Outreach and education for the Interagency OHH Research Program
- Rapid response capabilities

Much recent pathogen research has focused on advancements in methods that allow for rapid detection and source identification of harmful microbial organ isms in water and shellfish tissues. Traditional methods for detection are based on indicator bacteria rather than direct identification of enteric pathogens. These approaches are labor intensive, applicable only to a limited number of pathogens, and do not provide information about pollution sources. The OHH program has ongoing research aimed at detecting and determining the source of pathogens in recreational waters (whether from wildlife, domestic animals, or humans) accurately, rapidly, and inexpensively. Results from this work will greatly improve our ability to protect both public health and shellfish operations in coastal environments. Also, work supported by the OHH programs and others has led to a new understanding of soils and sediments, including beach sands, as reservoirs and sources of beach bacterial contamination.

Despite these advances, a number of scientific hurdles remain. Application of DNA-based techniques for rapid detection, quantification, and tracking is severely limited by the lack of suitable concentration methods for initial isolation of microbes from large volumes of water and other complex environmental samples. Targeted research in this area should enable rapid development and deployment of a wide range of microbial sensors as part of ocean observing systems. While the ability to forecast the need for beach and fishery warnings/closures has advanced, much more robust models are needed to evaluate the transport and fate of pathogens in aquatic systems and to understand the roles of sediments as reservoirs. Reliable prediction of the fate and transport of pathogens is essential to support decisions about beach closures, inform the design of structures for combined sewer out- falls and water intakes, and address soil and sediment as sources of pathogens.

OHH research should be aimed at how pathogenesis and virulence are linked to ecosystem variables, including seasonal variability in water temperature, water chemistry, algal blooms, climate cycles, and global climate change. For example, sea surface temperatures along the Gulf Coast have been found to account for over 50 percent of the variability in concentrations of disease- causing *Vibrio* bacteria. Also, some evidence connecting coastal environmental degradation to human health threats is emerging, such as an apparent link between metal contamination of coastal waters and proliferation of antibiotic-resistant bacteria, which in turn could pose additional risks to humans exposed to pathogenic bacteria in coastal waters. These kinds of associations require much further study. Improved disease surveillance would allow earlier detection of outbreaks, making it possible for public health officials to respond more rapidly to avoid or control them, mitigate exposures, and decrease the number of people who become ill.

Major opportunities for advancement related to waterborne pathogens require integration of OHH research and activities with epidemiology, public health, resource management, and risk communications (see Epidemiology and Outreach and Education sections) and include the following:

- Conduct basic and applied research about known and emerging human and marine animal pathogens, their transport and occurrence, including development and use of models and predictive and preventive capabilities related to disease trends.
- Determine how the prevalence and virulence of pathogens are linked to coastal and oceanic ecosystem variables, including seasonal variation, water temperature, water chemistry, algal blooms, climate cycles, and trends in global climate change.
- Improve molecular techniques to allow for rapid, direct, and simultaneous detection of multiple harmful marine organisms and incorporate into ocean observing systems.
- Improve techniques for concentrating microbes from large volumes of water and other environmental samples to allow more effective use of molecular and other modern pathogen detection methods.

- Develop standards, standard methods, and centralized culture collections, such as mixed cultures and other biological materials specifically relevant to OHH research (see Standards and Standardized Methods section).
- Develop risk assessment tools with the ability to integrate effects of multiple ecosystem stressors on pathogen presence, fate, and transport, and use sentinel species and wildlife disease surveillance to elucidate and predict potential human health outcomes.

BOX 13: NSF-NIEHS OHH CENTER STUDIES HUMAN PATHOGENS IN MT. HOPE BAY, MASSACHUSETTS

Mt. Hope Bay, Massachusetts has been heavily impacted by human activity, resulting in significant pollution due to thermal effluents, sewage outfalls, and the effects of past industrial activities. Scientists at the WHCOHH are sampling bay waters for the presence of a wide range of human pathogens and examining the potential role of thermal and sewage input on persistence of these pathogens. Their work is the first to document the presence and versity of *Legionellae*, the group of bacteria that cause Legionnaire's disease, in the marine environment. While many of the genetic sequences recovered are similar to those obtained from freshwater environments, they also detected the human pathogen (*L. pneumophila*) in Mt. Hope Bay samples. Microbial diversity assessed using genomic techniques revealed that the largest number of potential human pathogens were in sediment samples rather than in the water column.

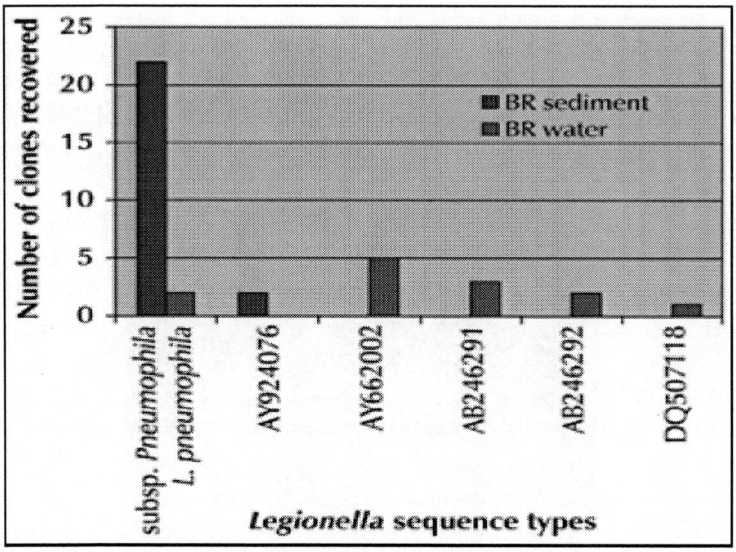

Diversity of the *Legionella* sequence types, known to cause Legionnaire's disease that were recovered in the water vs sediment in Mt. Hope Bay, Mass.. Photo credit: R. Gast.

Colonies of fungi (which may be pathogenic) are isolated from the upper respiratory tract (blowhole fluid) of wild bottlenose dolphins from Charleston, South Carolina and grown on this agar plate. Photo credit: P. Morris.

BOX 14: PATHOGENS DETECTED THROUGH GROUNDWATER QUALITY INVESTIGATION

A major tourist destination of Lake Erie, South Bass Island, also known as the "Key West of the Midwest," was the source of groundwater contamination that sickened 1,450 people during summer 2004. A scientific team at Michigan State University, part of the CEGLHH, sampled 16 drinking water wells on the island and found the groundwater contaminated with multiple fecal-associated pathogens. The CEGLHH developed and implemented a Lake Erie hydrodynamic model that showed a complex water movement pattern around the island preceding and during the time of the outbreaks. Analyses of model runs demonstrated that the massive contamination of drinking water wells was the result of heavy rains in May 2004 that contributed to higher groundwater levels, coupled with unique water movements that transported sewage and pathogens from public and private sewage treatment systems to drinking water wells. As a result of this investigation, the Ohio EPA and Department of Health are addressing the wastewater issue and supplying the island with fully treated drinking water.

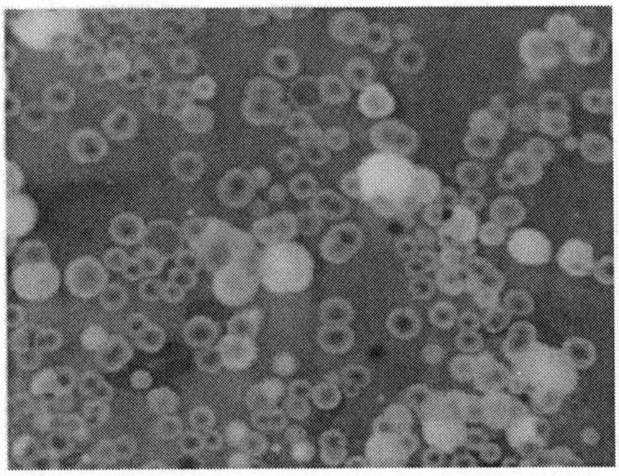

Vibrio bacteria are pathogens known to cause gastroenteritis and septicemia in humans from ingestion of undercooked or contaminated seafood. Colonies of *Vibrio parahaemolyticus* appear mauve color while *V. vulnificus* and *V. cholerae* appear turquoise blue on this agar plate. Photo credit: B. Robinson.

BOX 15: TRACKING PATHOGENS FROM SPACE: RISK ASSESSMENTS TO PREDICT SEAFOOD CONTAMINATION

Scientists at the University of Southern Mississippi, with support from NOAA's OHHI External Grants Program, are using sea surface temperature, chlorophyll, and turbidity data collected by satellite to develop and test a surveillance system for at least two disease-causing marine bacteria, *Vibrio parahaemolyticus* and *V. vulnificus*. Until now, these harmful marine organisms have been difficult to detect because sample collection is labor and resource intensive, making it difficult to predict potential outbreaks. People who eat raw or undercooked seafood (e.g., oysters) containing high levels of these naturally occurring bacteria or whose open wounds are exposed to seawater containing high levels of these bacteria may become ill. Symptoms range from mild gastrointestinal illness to more serious and sometimes lethal infections especially in immunocompromised persons. By harnessing the technological capabilities of satellites, in combination with application of molecular biological detection methods, it may be possible to better predict and prevent human disease associated with food-borne or recreational water exposures.

BOX 16: BEACH CONDITIONS "NOWCAST" FOR HUNTINGTON BEACH, LAKE ERIE

During summer 2006, the USGS and the Cuyahoga County Board of Health instituted and tested a system to quickly estimate bacteria levels and provide beach advisories to swimmers headed to Huntington Beach in Bay Village, Ohio. By 9:30 each morning, a Nowcast estimate of bacterial levels was posted, enabling the public access to advisory information before they left for the beach. Estimates are made using a computer model calibrated especially for Huntington Beach, which takes into account current weather and environmental conditions. Nowcast information for Huntington Beach is available on the Web at http://www.ohionow-cast.info/, and the general methodology is available for possible use at other beaches (http://pubs.water

Chemical Contaminants

Chemical contaminants are ubiquitous in aquatic environments and can pose risks to humans either via consumption of contaminated seafood or from contact with polluted waters. Urbanized estuarine watersheds are well documented to have elevated contaminant levels in their sediments, water, and associated biota. Until recently, the majority of research has focused on the effects of metals and persistent organic pollutants such as methyl mercury, pesticides, polychlorinated biphenyls (PCBs), polycyclic aromatic hydrocarbons (PAHs), and oil spills. EPA's Oil Spill Program, NOAA's Office of Response and Restoration, and the U.S. Coast Guard coordinate oil- and chemical-spill prevention and response capabilities in coastal waters. A number of agencies conduct contaminant monitoring programs in coastal areas, such as EPA's Environmental Monitoring and Assessment Program and National Coastal Assessment Program, NOAA's National Status and Trends Program, and USFWS National Wetlands Inventory, as well as Great Lakes information from the State of the Lakes Ecosystem Conference. Data collected from these monitoring efforts are used as the basis for the National Coastal Condition Report, a multi-agency effort led by the EPA to assess the condition of waters, sediments, and biota around the entire coastline of the United States.

Although successful programs monitor many of the chemical contaminants in our waters, many other potentially dangerous chemicals are not monitored routinely because of lack of methodologies for detection and quantification and/or lack of information on potential hazards posed to aquatic ecosystems and humans. Recent research has shown that very low levels of PAHs affect fish heart function, raising questions about potential for increased heart disease in people exposed to PAHs, perhaps even through air pollution (see Sentinel Species, Habitats, and Biological Models section). Also, studies are necessary to investigate the balance between the health benefits of eating seafood versus the risks associated with increased exposure to methyl mercury and other contaminants in seafood (see Seafood Safety section).

In recent years, OHH researchers have begun to investigate how increased levels of emerging chemical contaminants (ECCs), such as flame retardants, stain repellants and stick-resistant coatings, human and animal-use pharmaceuticals, and pesticides in aquatic environments, could affect ecosystem and human health. These ECCs are released into coastal watersheds through upland runoff from both urbanized and agricultural lands, sewage discharges, industrial releases, atmospheric transport, and aquaculture. Some ECCs have the potential to bioaccumulate and can be transferred throughout food webs. One group of ECCs that could be addressed through OHH research is the class of brominated flame retardants. Flame retardants have been reported to cause neurological effects in mammals and reduced growth and survival of juvenile fish. These compounds have not been monitored widely in estuaries, and little information is available regarding their toxicity in estuarine species. Additionally, the potential for transfer of these contaminants to humans consuming seafood products has not been well characterized, although humans are known to accumulate them.

Other ECCs of concern include perfluorinated compounds (PFCs) which contain chemicals with unique properties to make materials stain- and stick-resistant such as non-stick pan coatings, fabric stain repellants, microwave popcorn bags, pizza boxes, fire-fighting foam, and some cosmetics products. PFCs are resistant to breakdown and are turning up in unexpected places around the world, including humans and marine wildlife. Recent studies of sea turtles have shown PFCs surpassing PCBs in concentration. PFCs are persistent, bioaccumulate and can cause tumors and toxicities to the liver, reproductive, developmental, and immune systems of rats, mice, and monkeys. The effects on humans have not been observed or are inconclusive.

Chemical contaminants which stem from industrial processes, household activities and vehicles are released into coastal watersheds through sewage outfalls, runoff and other point and non-point discharges.

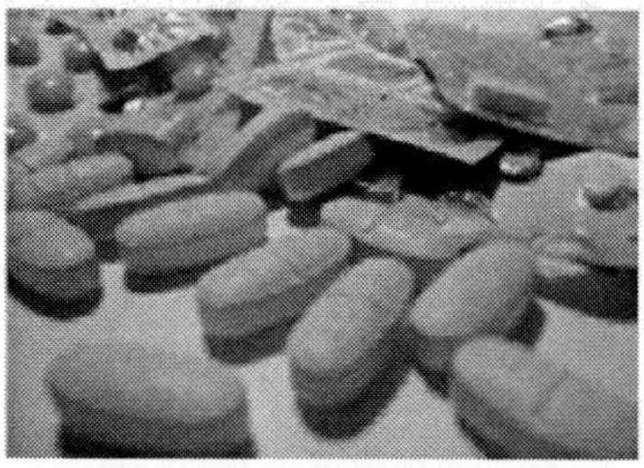

The effects of pharmaceuticals and personal care products in the marine environment are less studied compared to freshwater but already show similar effects such as the feminization of male marine fish.

The increasing prevalence of pharmaceuticals in aquatic ecosystems could have widespread effects on marine organisms because pharmaceuticals are biologically active and may cause significant unintended responses even in minute quantities (e.g., endocrine-disrupting reproductive impacts). These compounds also may be transferred to humans through consumption of fishery products. In addition, the release of antibiotics to the environment may favor the development of resistant strains of pathogenic bacteria or lead to alterations in the composition of natural microbial communities, which in turn may have severe consequences for both human and ecosystem health.

Aditional ECCs that warrant further research include a wide variety of insecticides, fungicides, and herbicides used to control insect, weed, and fungal pests associated with crops, turf grass, and ornamental plants and new anti-fouling compounds used to limit marine growth on boat hulls (see Box 17). Although modern pesticides are typically less persistent than earlier classes of pesticides, many are acutely toxic to a variety of non-target estuarine species. Because of the way they are applied, they may enter the environment in the form of repetitive pulses at concentrations sufficient to cause toxicity in non- target species. New compounds enter the market every year and their toxicity to most estuarine species is often unknown. Thus, analytical, toxicological, and ecological methodologies to document and quantify their concentrations and potential effects in estuaries are essential.

The number and diversity of both common chemical contaminants and ECCs that may impact human or coastal ecosystem health are tremendous and represent an important focus of the Interagency OHH Program. Major opportunities to advance understanding of chemical contaminants in coastal waters include the following:

- Conduct basic research and monitoring to provide baseline information to identify and prioritize known and emerging chemical contaminants.
- Develop, validate, and standardize analytical methods for the detection of these contaminants in estuarine and marine waters, sediments, and tissue samples (see also Standards and Standardized Methods section).
- Develop sensitive indicators (biomarkers) of ECC effects in estuarine and marine environments.
- Regularly monitor estuarine and marine ecosystems, seafood, and humans to determine the presence and distribution of ECCs.
- Evaluate ECC hazards to estuarine and marine biota and especially to humans who may be exposed via multiple pathways.

- Develop improved models and risk assessments to inform management and public health authorities.

Harmful Algal Blooms

Blooms of various kinds of algae occur commonly and naturally in coastal waters, estuaries, lakes, and ponds. Some, typically referred to as HABs, have harmful effects or produce potent natural toxins that cause mild to severe disease and even death among a variety of aquatic species and humans exposed to them *via* consumption of contaminated seafood or other avenues (Table 1). Several reports highlight global increases in frequency and intensity of HABs, leading to more closures of beaches and fishing grounds and associated economic losses. Recent work with dolphins and manatees has documented that some deaths of these marine mammals occurred as a result of their consumption of toxin-contaminated fish and seagrass, raising the specter of potentially increased dangers to human and animal consumers of seafood. Also, OHH research has documented an association of pathogenic bacteria with HABs.

Box 17: NOAA's OHHI Develops New Methods to Identify and Measure Toxicity of Antifouling Pesticides

Many new antifouling agents have been created and widely used to prevent settlement and growth of marine organisms on submerged surfaces of vessels. While such compounds have clear commercial benefits, numerous questions concerning unintended impacts of new antifouling agents in ocean environments have been raised. OHH researchers at the HML developed a sensitive liquid chromatography-mass spectrometry method to detect several currently used antifouling agents. Monitoring studies were initiated in several locations including South Carolina, Puerto Rico, and California. A parallel laboratory evaluation of the potential toxicity of one such agent, Irgarol 1051, to several key estuarine organisms, including grass shrimp and clams, indicated that these animals were relatively unaffected by the chemical. However, algae were found to be very sensitive to Irgarol 1051 at concentrations similar to those measured in the environment during the monitoring studies. Impacts to algae at the base of estuarine food chains could seriously affect higher-level consumers.

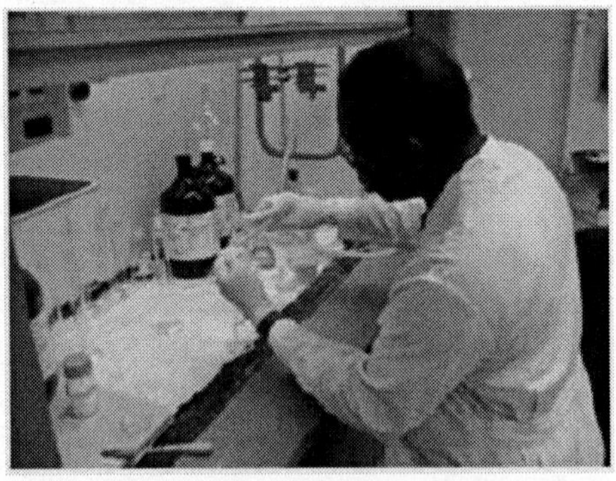

Scientists study the toxicity of chemicals on various marine species, including pesticides and anti-fouling compounds painted on boats. Photo credit: S. Lovelace.

On the other hand, some HAB toxins may hold promise for development of new pharmaceuticals or other useful products (see Pharmaceuticals and Other Beneficial Products section).

For over a decade, the National Plan for Marine Biotoxins and Harmful Algae has guided U.S. HAB research and has led to considerable progress in understanding blooms and their toxins. In 2005, a new 10-year research plan (*HARRNESS, Harmful Algal Research and Response: A National Environmental Science Strategy 2005–2015*) identified four broad areas for focused research: (1) bloom ecology and dynamics, (2) toxins and their effects, (3) food webs and fisheries, and (4) public health and socioeconomic impacts. An associated report, *Harmful Algal Research and Response: A Human Dimension Strategy*, outlines some of the social, behavioral, and economic studies needed, and the *National Assessment of Efforts to Predict and Respond to Harmful Algal Blooms in U.S. Waters*, provides recommendations dealing with prediction and control of HAB events.

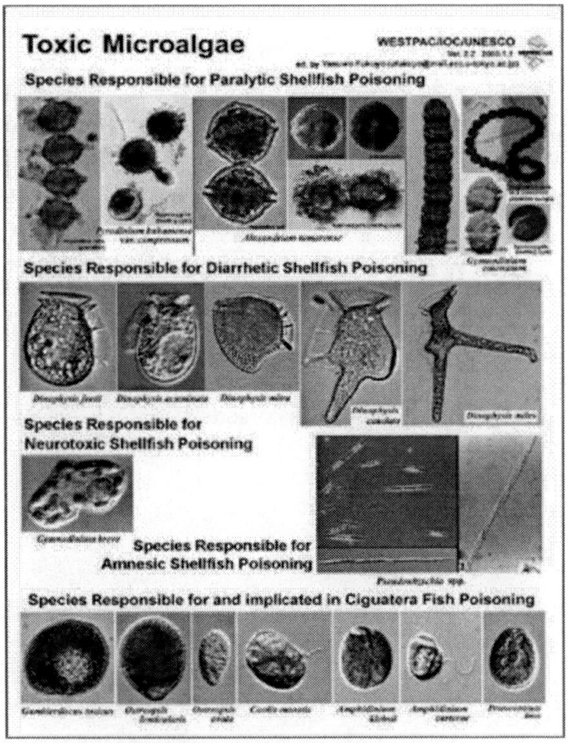

Many species of algae are responsible for producing harmful algal bloom toxins. Photo credit: With permission from TOS and source: http://dinos.anesc.u-tokyo.ac.jp/Jpeg/index.htm.

Prior to existence of the Interagency OHH Program, Congress passed the Harmful Algal Bloom and Hypoxia Research and Control Act of 1998 (amended in 2004) and put into motion an Interagency HAB research effort. Thus, the Interagency OHH Program will complement the ongoing federal agency HAB activities, with primary, although not exclusive, focus on documenting and understanding direct and indirect HAB-related health threats to humans and marine animals, collaborating to conduct epidemiological studies, new sensor technologies, and opportunities for beneficial uses of HAB toxins (see Boxes 18-20).

Major opportunities for advancement within the Interagency OHH Program include the following:

- Develop, test, deploy, and validate new or improved sensors for rapid, accurate, and inexpensive identification and quantification of multiple HAB species and HAB toxins in the environment, with particular emphasis on incorporating such sensors into ocean observing systems.
- Increase accuracy and timeliness of estimates of the concentration and distribution of waterborne harmful algae and toxins. Identify more accurate indicators of risk to humans and animals and improve measurement techniques to reduce the time between sample collection and the availability of results.
- Collaborate to develop and test new models that combine health and environmental surveillance data to produce useful forecasts and risk assessments for HAB-related health impacts and assess the capabilities of the ocean observing systems to support public health decision-making.
- Improve understanding of modes of action and health effects of HAB toxins in animals and humans to develop better health warnings and treatments.
- Investigate the occurrence and significance of observed associations of pathogens such as bacteria and viruses with HABs in ecosystem and human health contexts (see also Pathogens section).

BOX 18: NOAA'S OHHI DEVELOPS PCR ASSAY TO DETERMINE TOXICITY OF *MICROCYSTIS* BLOOMS IN THE GREAT LAKES

Microcystis is a particularly harmful alga in the Great Lakes. Much more information is needed about factors that trigger blooms of this organism and when such blooms become toxic and thus pose risks to humans. Researchers at NOAA's CEGLHH developed a biochemical assay to quickly identify whether colonies of *Microcystis* were capable of producing toxins and to help identify which blooms are particularly threatening to recreational beaches and drinking water sources in the Great Lakes.

Microcystis harmful algal bloom in western Lake Erie and the infamous Microcystis "shake" taken from a water sample. Photo credit: T. Bridgeman.

Seafood Safety

The health benefits of eating seafood are well known. Fish are an important source of high quality protein and other essential nutrients and contain omega-3 fatty acids, which have a variety of benefits. Yet seafood is also known to pose some health risks to humans if it is contaminated with pathogens or biotoxins, or accumulated heavy metals, organic pollutants, and ECCs.

BOX 19: NOAA'S OHHI TRACKS HABS ON THE WEST COAST

NOAA's WCCOHH, in collaboration with the ECOHAB Pacific Northwest Program, deployed drifters in the Juan de Fuca eddy, a known HAB "hot spot" off the Washington coast. These drifters track the movements of the toxic patches, providing an early warning of need for coast-wide closures of recreational, commercial, and tribal subsistence razor clam fisheries.

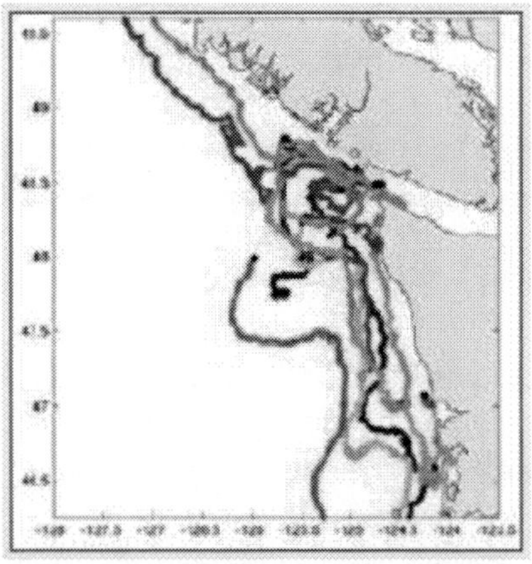

Drifters deployed during the September 2005 ECOHAB Pacific North - west cruise track movement of harmful algal patches as seen here in the Juan de Fuca eddy region. Photo credit: V. Trainer.

NOAA, with additional support provided by the FDA, asked the Institute of Medicine of the National Academies (IOM) to review evidence on the benefits and risks associated with seafood consumption. The IOM released its findings in October 2006 and confirmed seafood is rich in nutrients, low in saturated fats, and should be incorporated more into the American diet to reduce the risk of early death by heart disease and stroke. Furthermore, the IOM found pregnant and nursing mothers and children should consume some seafood because of the significant benefit to cognitive development and association with higher IQ in newborns and children. At the same time as the release of the IOM study, a meta- analysis by Harvard School of Public Health researchers found reductions in death rate from heart disease among those who ate at least one to two servings of fish per week. A significant opportunity for the Interagency OHH Program is to determine the full range of health benefits from fish consumption, such as learning how much fish must be consumed to achieve these benefits, the active components in fish that protect humans from disease, and the mechanisms by which such protection occurs.

Contrasting with the benefits of fish consumption are risks associated with contamination by pathogens, toxins, and chemical pollutants (see Box 21). For example, contamination of fish with mercury and the potentially harmful effects of mercury on brain development pose a significant problem in communicating the balance of risks and benefits to the public and especially to women of child-bearing age. Pregnant women and nursing mothers should consume some fish to realize the benefits, but they should not eat certain fish species, which contain high levels of mercury. One opportunity for the OHH Interagency Program is to assist in standardizing risk and benefit analysis across agencies to ensure that recommendations and health advisories do not provide confusing or conflicting information to the public.

Box 20: Bivalves and Food Web Transfer of Domoic Acid in Washington Waters

Laboratory studies are being conducted at the NSF-NIEHS Center for Oceans and Human Health at the University of Washington (PNW H2O) to determine how various species of harvestable shellfish (including local species of clams, oysters, and mussels) differ in the rates they feed on the toxic alga, *Pseudo-n itzschia*. The feeding rate may influence the amount of domoic acid toxin retained in the shellfish tissue. Preliminary results suggest that mussels and oysters consume *Pseudonitzschia* at higher rates than clams found in Puget Sound, and at substantially higher rates than razor clams found along Washington's coast. It has been shown previously that mussels purge domoic acid from their tissues quickly, whereas razor clams take a long time to purge the toxin. The high feeding rate of mussels and oysters found in Puget Sound, combined with the ability of mussels to quickly purge toxin from their tissues, may be related to the decreased frequency of beach closures for *Pseudo-nitzschia* in Puget Sound. Data from these experiments will be combined with a model currently being developed to track the offshore to onshore water movement to predict the impacts of *Pseudo-n itzschia* blooms on the intertidal shellfish in Puget Sound.

Feeding rates of shellfish such as clams might affect the amount of domoic acid toxin retained in the tissues. Photo credit: E. Dusek.

FDA, EPA and many states communicate the risk of eating certain fish high in mercury content to pregnant and nursing women, children and the general public. Source: http://www.epa.gov/mercury/advisories.htm.

The need for risk and benefit analysis of seafood also applies to aquacultured products, as the majority of U.S. seafood comes from international sources, with nearly half the imports from aquaculture. The increasing proportion of aquacultured fish raises additional concerns about the amount of beneficial omega-3 fatty acids in the fish versus the chemical contaminant loads stemming from the feeds used to grow fish. Research is needed to identify and develop more plant- or microbe- based components to replace meal and oil in aquaculture feeds. Such replacement could address concerns related to both the sustainability of the wild fish stocks exploited for production of meal and oil and the levels of chemical contaminants these products may contribute to aquaculture feeds. Current research within NOAA's OHHI is developing technologies to produce competitively priced aquacultured seafood with enhanced beneficial fatty acid profiles and safe contaminant levels. Improving the nutritional characteristics of aquacultured seafood is an area where enhanced research within the Interagency OHH Program could result in major benefits for the public and a competitive edge for U.S. aquaculture producers in the global seafood market.

Development of better detection methods and forecasting capabilities for pathogens and toxins are underway (see Box 22), as are new approaches to destroy or make harmless pathogens and toxins in seafood and to educate the public about seafood consumption. Such capabilities are being developed within the Interagency OHH Program and can be addressed through targeted research, technology development, and implementation of monitoring protocols combined with new technology and communication tools. In addition, the National Aquatic Animal Health Plan, a cooperative effort of the Animal and Plant Health Inspection Service of the U.S. Department of Agriculture, NOAA Fisheries of the U.S. Department of Commerce, and the USFWS of the U.S. Department of the Interior working together under the Joint Subcommittee on Aquaculture, is being developed to protect cultured and wild aquatic animal products from disease and ensure availability of diagnostic and certification services.

Although some benefits and risks of consuming fish are well established, continued research is needed to more fully understand the human health benefits and risks of seafood consumption and develop better approaches to ensure the high quality, safety, and sustainability of the nation's seafood supply.

Several major opportunities for advancement that could produce significant health benefits and reduced health risks for American seafood consumers include the following:

- Cooperate with the National Aquatic Animal Health Plan and Task Force to focus on major animal health research topics of mutual interest.
- Expand development of assays, sensors, microarrays, and other techniques to detect and quantify levels of specific contaminants of concern (pathogens, harmful algal toxins, chemical pollutants) in seafood from all sources, including marine wildlife harvested by subsistence users, that can be applied in the field and used by industry (e.g., fishers, fish markets, fish wholesalers and restaurants, etc.) and public health workers (see Pathogens, Chemical Contaminants and HABs sections).
- Examine seafood safety in an ecosystem context, including the role of anthropogenic factors, climate cycles, and climate change in pathogenesis, toxicity, and contamination of seafood, and develop and implement models linking environmental data to seafood contamination to build predictive and risk assessment capabilities.
- Determine the full range of health benefits from fish consumption, coupled with focused studies on which seafood components provide benefits and the biological mechanisms of those benefits. Specific research targets might include determining preferred levels and ratios of particular fatty acids in fish flesh and to what extent the selenium content of fish flesh may offset potentially harmful effects of mercury contamination.
- Standardize comparative risk assessments that consider both beneficial and negative characteristics of specific types or sources of seafood in order to reconcile recommendations and health advisories and effectively communicate to the public.
- Evaluate new ways to enhance the levels of beneficial omega-3 fatty acids and reduce chemical contamination in aquacultured fish and fish feeds, including the use of materials derived from plants and microbes to substitute for fish meal and oil as basic components of fish feeds.

Box 21: NOAA's OHHI Investigates Benefits and Risks of Seafood Consumption

NOAA-supported researchers at the HML measured the chemical contaminant levels and fatty-acid content of samples of red drum (a popular marine fish) from a range of domestic and imported sources, including both wild and cultured fish. The results, interpreted in terms of both human health risks and benefits, suggest that fatty acid ratios and contaminant loads vary widely according to source and in some instances may pose a health hazard. A similar investigation of shrimp from 62 domestic and foreign sources is currently underway. HML scientists are using information derived from these studies to develop and test aquaculture technologies that will permit economically competitive production of seafood with certifiably safe contaminant levels and enhanced, beneficial fatty acid profiles.

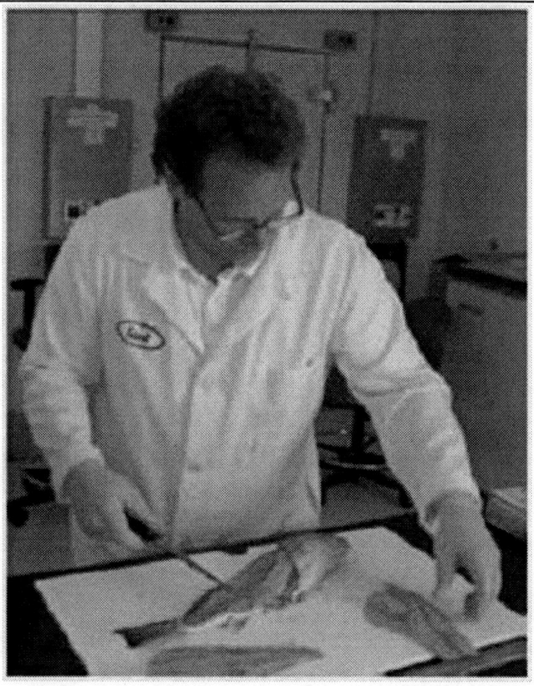

Red drum are harvested from wild populations, and are also farmed. Scientists are analyzing fish tissues to see how contaminant loads vary by source. Photo credit: S. Lovelace.

Box 22: NSF-NIEHS Ciguatera Studies

Ciguatera poisoning is a HAB-related illness caused by the consumption of fish, usually more tropical species, that have become contaminated with ciguatoxins produced by a dinoflagellate, *Gambierdiscus spp.* Research conducted by OHH researchers at the PRCMB is pursuing improved strategies to predict ciguatera outbreaks and prevent human poisonings arising from the consumption of ciguatoxic fish. Ecological efforts include the collection and analyses of fish, algae, and environmental data from coastal sites throughout the Hawaiian archipelago, and physiological studies that focus on the identification of conditions that trigger and promote the production of ciguatoxins by the dinoflagellate. Work is also underway to improve analytical methods to detect ciguatoxin at the extremely low levels that are known to cause poisonings in humans. Analyses have shown that between 12–17 percent of the fish sampled from Hawaiian coasts are ciguatoxic. Collaborative research has identified a novel ciguatoxin congener in Hawaiian fishes named CTX-H. The isolation and purification of this material is being used to develop improved detection methodologies needed to prevent ciguatoxic fish from reaching the consumer. In addition to *Gambierdiscus toxicus*, the species of dinoflagellate thought to produce ciguatoxin, two entirely new species of *Gambierdiscus* have tentatively been identified in Hawaiian waters.

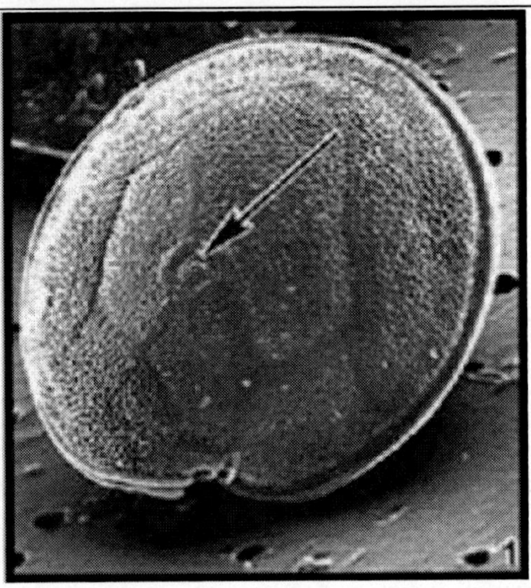

Gambierdiscustoxicus (shown here) produces ciguatoxins which can accumulate in tropical reef fish via the food web and can cause illness in seafood consumers. Source: http://www.nmnh.si.edu/botany/projects/donoglag/Taxa/ Gtoxicus.htm.

Pharmaceuticals and Other Beneficial Products

For thousands of years, humans have used natural products from the land to treat diseases. Over time, humans learned how to isolate bioactive and other useful chemical components from plants and microorganisms, particularly bacteria and fungi. These discoveries eventually led to synthesis and mass production of purified medicines and other beneficial products. The discovery of penicillin in the late 1920s was arguably the single most important medical advance of the 20th century, and altogether over 120 antibiotics and other drugs originally derived from soil microbes are still in use today. Yet, despite a growing problem of antibiotic resistance, the global drug industry is not emphasizing continued exploration for and development of a new generation of safe and effective antibiotics for human use.

The potential for ocean resources to provide new sources of drugs and beneficial products, including some perhaps as transformational as the discovery of penicillin, is just beginning to be widely appreciated. Pioneering research began as early as the 1960s, including efforts supported by NOAA's National Sea Grant College Program, NIH's National Cancer Institute, the NSF, and other agencies, and some truly remarkable discoveries have been made (Table 2). However, major programs focused on discovery and development of drugs from the sea have not materialized. Recent advances in molecular biology, high-throughput cultivation of microorganisms, growing knowledge and exploration of deep-sea environments and chemosynthetic ecologies, and understanding of the ocean's incredible biodiversity, along with mounting problems of drug-resistant pathogen strains and increased incidences of some kinds of cancers, have resulted in new interest in the oceans from a pharmaceutical/natural products standpoint. For example, spongouridine and spongothymidine, compounds derived from a sea sponge, prompted the development of the synthetic anti-leukemia drug cytarabine (Cytostar®), and two very important anti-viral drugs

(Acyclovir and AZT, which are used to treat Herpes and HIV infections, respectively, were based on the same natural marine sponge compounds.

A veritable host of marine species have demonstrated potential to yield useful compounds. These include microorganisms, principally the bacteria and fungi, but also algae, corals, mollusks, and fish. Several of these are already in clinical trials to treat cancers, tumors, inflammation, Alzheimer's disease and for use with bone implants, cosmetics and food additives (Table 2). Limited but promising studies indicate that some HAB toxins may have potential to treat cancer and pain in humans or may be used broadly as chemical tools for biomedical research. Also, there is growing private-sector interest in working with novel-disease organisms isolated from marine mammals and other marine animals to determine their potential for use in developing new compounds for human applications (see Box 23).

The advent of new, extremely sensitive analytical capabilities (including powerful nuclear magnetic resonance and mass spectral instruments) has opened the field to the discovery of highly active metabolites that exist in such low concentrations that they were previously beyond reach. In addition, the ability to mass culture toxin-producing algae, bacteria, and fungi under highly controlled and safe conditions allows for production of sufficient quantities of toxin for experimentation and testing. The availability of appropriately equipped biosafety laboratories opens new opportunities for beneficial advancements in the discovery and development of therapeutics based on materials from the sea. Coupling these laboratories with new remotely operated and in situ devices that allow sampling of very small quantities of biological materials from extreme and novel environments, without risk of introducing pathogens or invasive species, offers great potential for new discoveries.

Major opportunities for advancement include the following:

- Examine genetic, species, and molecular diversity within extreme marine habitats (e.g., thermal vents, seamounts, deep sediments, and cold seeps) using ecologically sound methods and evaluate these environments as sources of bioactive compounds.
- Develop improved mass-culture methods to allow large-scale production of marine microbes, HAB species, and other organisms that produce compounds of interest for the development of pharmaceuticals and biochemical probes.
- Conduct focused studies to understand the chemical structures and functional relationships between algal toxins' toxicities and their effects, and these toxins' potential as beneficial products, such as new therapeutics and reagents (see HAB section).
- Investigate novel microbial diversity associated with corals and other organisms, including understanding their mechanisms of chemical "warfare" and defense, as well as symbiotic interactions that may result in the generation of bioactive metabolites/chemicals with potential utility in development of drugs and other beneficial products.
- Evaluate more thoroughly the widespread distribution of marine-based neurotoxins and their possible links to health of humans, fish, and birds; mechanisms of action; and potential applications to treat or prevent disease (see HAB and Epidemiology sections).
- Make available methods and materials to rapidly and accurately screen newly isolated compounds for bioactivity (e.g., ability to affect tumors, inhibit bacterial growth) and develop more sensitive and selective bioactivity-screening approaches.

- Establish mechanisms to archive portions of bioprospecting samples (including DNA for subsequent metagenomic studies) in order to conserve samples collected with federal funds as national resources (see also Standards section).

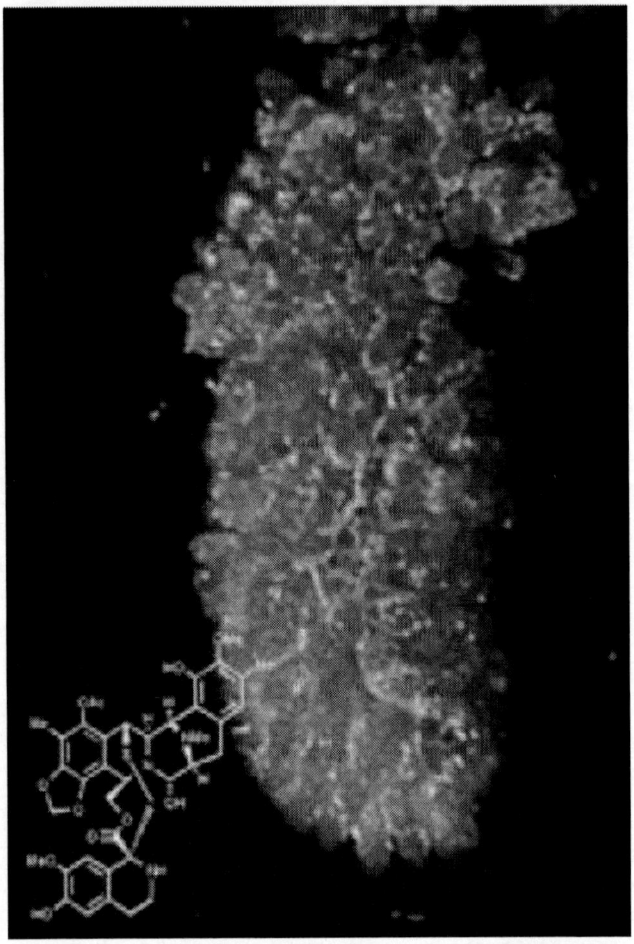

- The sea squirt *Ecteinascidia turbinata* grows on mangrove roots and offers powerful anti-cancer compounds. Photo credit: With permission from The *Oceanography* Society.

- Enhance collaborations and partnering among the marine biological, oceanographic, and biomedical communities and with the private sector.
- Access vessels and other platforms for ecologically sound exploration and sampling of poorly known marine environments and at locations where the potential for discovery of promising compounds is high (see "Access to Sea").
- Establish long-term mechanisms (e.g., the NOAA and NSF-NIEHS OHH Centers) to support drug and product discovery demonstration efforts over the extended time periods necessary to move from the ocean to the laboratory to commercial products.
- Implement consistent intellectual property procedures across federal agencies involved in OHH research.

BOX 23: PHARMACEUTICALS FROM THE SEA

In 2004, Diversa Corporation signed a biodiversity access and collaboration agreement with the University of Hawaii to cultivate novel bacteria associated with marine animals, sea grasses, and ocean sediments from Hawaiian samples. This discovery effort was co-funded by NOAA's OHHI and the NSF-N IEHS OHH Center at the University. This collaborative agreement provided a unique opportunity to integrate Diversa Corporation's high throughput culturing (HTC) technology with Hawaii's unique biodiversity and expertise in marine natural products chemistry. Samples of the sponge *Mycale armata* were collected from Kaneohe Bay, off Oahu, along with sediments sampled from various locations. Screening results revealed that five of the HTC bacterial isolates obtained from the sponge exhibited anti-microbial activity and four had anti-cancer activity. These findings indicate this strategy has great potential for discovering new drugs from the sea.

Bacteria associated with the sponge *Mycale armata* have been discovered by OHH scientists to exhibit anti-microbial and anti-cancer activity. Photo credit: S. Coles.

Table 2. Status of Marine-Derived Natural Products in Clinical and Preclinical Trials

Compound	Source	Disease	Comment
Bryostatin 1	Bryozoan (a sessile marine animal, also called moss animal) *Bugula neritina*	Cancer	Now in combination therapy trials; licensed to GPC Biotech by Arizona State University
TZT-1 027	Synthetic Dolastatin isolated from the sea hare (a type of sea slug)	Cancer	Also known as Auristatin PE and Soblidotin; in trials
Cematodin	Synthetic derivative of Dolastatin 15 iso- lated from the sea hare	Cancer	Some positive effects in melanoma; in trials
ILX 651, Synthatodin	Synthetic derivative of Dolastatin 15 isolat- ed from the sea hare (a type of sea slug)	Cancer	For melanoma, breast, and non-small cell lung cancer
Ecteinascidin 743	Ascidian *Ecteinascidia turbinate* (mangrove tunicate)	Cancer	Licensed to Ortho Biotech (J&J/Jansen Pharmaceuticals); in trials
Aplidine	Ascidian (sea squirt) *Aplidium albicans* (Mediterranean tunicate)	Cancer	Dehydrodidemnin B; made by total synthesis; in trials
E7389	Marine sponge *Lissodendoryx sp.*	Cancer	Eisai's synthetic halichondrin B derivative; breast and lung; in trials
Discodermolide	Deep-water marine sponge *Discodermia dissoluta*	Cancer	Licensed to Novartis by Harbor Branch Oceanographic Institution; in trials
Kahalalide F	A marine invertebrate, Mollusk *Eylsia rufescens* and the feather haired alga *Bryopsis sp.*	Cancer	Licensed to PharmaMar by University of Hawaii; in trials
Zalypsis	Synthetic safracin B; derivative from marine bacteria	Cancer	PharmaMar (based on saframyin molecule); in trials
ES-285	Stimpson's surf clam, *Spisula polynyma*	Cancer	A protein involved with functioning of the nucleotide guanosine triphosphate important in metabolism; in trials
KRN-7000	an Okinawan sea sponge, *Agelas mauritia*-Cancer *nus*	Cancer	An agelasphin derivative; in trials
Squalamine	Spiny dogfish (shark) *Squalus acanthias*	Cancer	Anti-angiogenic activity; in trials
Æ- 941 (Neovastat)	Shark (spiny dogfish and hammerhead shark) cartilage	Cancer	Defined mixture of < 500kDa from cartilage; anti-angiogenic; in trials

Table 2. (Continued)

Compound	Source	Disease	Comment
NVP-LAQ824	Synthetic derived from a marine sponge	Cancer	Derived from psammaplin, trichostatin and trapoxin structures; in trials
E-7974 (Eisai)	Marine natural product hemiasterlin syn- thetic derived from marine sponges	Cancer	Carboxylate-end modified hemisasterlin; in trials
Salinosporamide A (NPI-0052)	Marine bacterium *Salinispora tropic*a	Cancer	Proteasome inhibitor Nereus Pharma; in trials
GTS-21 (aka DMBX)	Marine worm	Alzheimer's	Licensed to Taiho by the Univeristy of Florida; in trials
I PL-576,092 (aka HMR-4011A)	Marine sponge *Petrosia contignata*	Anti-asthmatic	Derivative of contignasterol; Inflazyme Pharma; in trials
I PL-512,602	Derivative of 576092 from marine sponge *Petrosia contignata*	Anti-asthmatic	With Aventis. No further data as of 8/2005
I PL-550,260	Derivative of 576092 from marine sponge *Petrosia contignata*	Anti-asthmatic	With Aventis. No further data as of 8/2005
Ziconotide (aka Prialt)	Cone shell species (Mollusk) from the Pacific Ocean *Conus magus*	Neuropathic pain	Licensed by Elan to Warner Lambert; approved by FDA 12/28/04; launched in U.S. and Europe in 2005
CGX-1160	Cone shell species (Mollusk) also known as 'anbonia' from Pacific Ocean C*onus geographus*	Pain	Cognetix and Elan Corporation (Ireland); in trials
ACV1	Cone shell species (Mollusk) also known as Queen Victoria Cone from Pacific Ocean *Conus victoriae*	Pain	Metbolic Pharma (Australia)(06/2006), conotoxin Vc1.1; in trials

(Adapted from Fenical, 2006. Oceanography 19(2):110-119).

Cross-Cutting Areas

Epidemiology

Epidemiologic research is a valuable tool to assess direct positive and negative impacts to human and animal health stemming from the oceans, coasts, and inland waters. Epidemiologic studies address the human part of the OHH equation by investigating diseases in human and animal populations most likely to be exposed to, or affected by, water-borne pathogens, and by observing and assessing naturally occurring disease patterns. The goals of epidemiologic research are to define the underlying characteristics of populations that make them susceptible to specific exposures and measure the associations between these population characteristics and environmental exposures. The results from epidemiologic studies can be used to assess cause and effect by examining the exposure-disease association, including its strength, consistency among studies, and whether or not the association makes sense from a biological standpoint (see Box 24).

A number of acute and chronic illnesses affecting humans and other animals are caused by ocean-based threats, such as infectious diseases and poisonings associated with chemical contaminants or algal toxins. Extreme coastal weather events, such as hurricanes, can put large populations at risk for exposure to harmful microbes, injury, and possibly death. Environmental epidemiology can help refine our knowledge of these public health threats, allow us to uncover emerging threats as they appear, and, ultimately, prevent human exposures to ocean-borne threats before they can cause disease.

Populations of interest to epidemiologists focused on OHH issues include those involved in seafood harvesting and handling, seafood consumers, environmental workers, individuals who work and play on or near the water, and coastal communities, especially indigenous peoples who rely on seafood for a substantial proportion of their diet. Other populations that may be particularly at risk from ocean-related exposures include children and people with compromised immune systems. In addition to directly assessing these populations of interest, aquatic animal health events can be evaluated to identify potential new human health threats (see Sentinel Species section).

Environmental epidemiology requires cross-disciplinary collaboration with experts from fields such as medicine, public health, epidemiology, toxicology, experimental design, analysis of environmental and clinical samples/ specimens, data management, and statistical analysis (see Boxes 25, 26). Public health surveillance is the systematic collection of health data that are then used to generate hypotheses about causal relationships, assess disease trends, and serve as the basis for public health protection activities. Of particular interest to the Interagency OHH Program, the CDC has developed HABISS a HAB-illness surveillance system (see Box 27) that can be used as a model for examination of other environmental health threats. As pointed out in the Ocean Research Priorities Plan and Implementation Strategy, "Ocean data and modeling should be integrated with assessments and epidemiologic studies to define exposures and focus risk assessments." One way to do this would be to build upon HABISS with participation by multiple agencies.

Opportunities to advance understanding of the scope of public health threats posed by the ocean and to improve public health protection include:

- Implement and enhance epidemiologic studies to characterize public health risks from pathogens, toxins, and chemical contaminants from the oceans and coastal waters (see Pathogens, Contaminants, and HABs sections).
- Characterize diseases associated with ocean-borne risks to improve diagnosis and treatment, reporting, and public health surveillance.
- Acquire, integrate, and interpret environmental data applicable to epidemiologic studies and disease surveillance systems at the right temporal and spatial scales and correlate with epidemiologic data (see Ocean Observing Systems section).
- Make improvements to disease surveillance to detect emerging human and animal health issues.
- Continue to support CDC's existing HAB-related Illness Surveillance System, including involvement of local health practitioners to the greatest extent feasible (see HAB section).
- Use epidemiologic data to help design and implement models to track and predict ocean-driven health events.
- Improve the transfer and applicability of relevant medical information from animal health studies and response investigations to the public health community.
- Conduct epidemiologic studies on direct and indirect effects of climate variability on the geographic and temporal distribution of ocean-borne human and animal health threats.

Box 24: Studies of Human Exposure to Domoic Acid

The Pacific Northwest is a unique location for the study of exposure to domoic acid because there are many diverse regional populations that consume a lot of shellfish. At the NSF-N IEHS Center for OHH at the University of Washington (PNW H2O), investigators have partnered with Native American populations who reside, harvest and consume shellfish in the PNW where domoic acid concentrations in razor clams, crabs and other shellfish have closed usual and accustomed areas of harvest. The University of Washington Center has also collaborated with the Washington State Department of Health and other partners to characterize dietary consumption, including specific information about shellfish, among Asian and Pacific Islander populations whose diets are traditionally high in locally harvested shellfish.

Subsistence and recreational harvesting of razor clams provides an important source of protein and socio-cultural activity for communities. Photo credit: C. McKay.

Box 25: NSF-NIEHS Center and CDC Epidemiologic Studies Related to HABs

The NSF-N IEHS Center for OHH at the University of Miami is collaborating with CDC, the Florida Department of Health, other universities, and private research laboratories to conduct epidemiologic studies of people exposed to aerosolized brevetoxins during Florida red tide events. The collaborative research program includes using animal models of asthma to provide a new level of understanding of the biological underpinnings of the effects observed in people. Study results have provided the basis for a further collaboration with NOAA to use human data to predict the effects of Florida red tides on coastal communities. Understanding the entire process from formation of the aerosols to the physiologic basis of the induced health effects has provided the foundation for an effective community outreach and education programs to prevent exposures and mitigate the public health impact from Florida red tide events.

Lifeguards are probably one of the populations most highly exposed to aerolized brevetoxins (*K. brevis*) because of their lengthy stays on beaches; some wear masks during harmful algal blooms. Photo credit: L. Backer.

Box 26: Multi-Agency Epidemiologic Studies Related to Water-Borne Illnesses

EPA is collaborating with the CDC and the USGS to conduct a series of epidemiologic studies to assess the occurrence of water-borne illnesses among beachgoers. In 2003 and 2004, the research focused on Great Lakes beaches and was to expand to Gulf of Mexico beaches in 2005, but this aspect was interrupted by Hurricane Katrina. Current plans are to conduct epidemiologic investigations along Atlantic and Pacific coastal beaches in collaboration with NOAA. For each study, nearly 8,000 volunteers report on their activities and health status after visiting the beach and EPA collects and analyzes water samples to assess exposure to water-borne pathogens. In addition, beginning in summer 2007, NOAA's OHHI will participate in a recreational waters epidemiological investigation organized by the Southern California Coastal Water Research Project in collaboration with an epidemiologist at the University of California, Berkeley. This partnership represents a rare opportunity to apply novel pathogen detection technology developed under the OHHI to water samples that will be collected simultaneously with health data.

Sentinel Species/Habitats and Animal Models

A variety of marine species and habitats are excellent indicators or sentinels of environmental stress and potential health threats for humans; they provide information about what is happening in the environment that may affect humans and animals. In-situ studies of sentinel species and habitats provide a means to assess cumulative or integrative effects of chemical contaminants, microbes, or toxins in food webs, animal systems, or ecosystems. Such sentinel species and habitats may be considered in at least three categories: (1) wildlife or habitats that tend to concentrate contaminants, toxins, and/or pathogens from their environment and thus may provide more biologically relevant indicators of exposure or effects than water sampling alone (e.g., filter-feeding bivalve mollusks such as oysters, clams, mussels); (2) wildlife with diets and/or physiologies at least partially similar to those of humans may demonstrate early indications of potential health effects (e.g., marine mammals, such as the California sea lion, bottlenose dolphins, and sea birds) (see Box 28); and (3) habitats encompassing key ecosystem components and subjected to high pollution exposure, thereby indicating potential effects at systems or community levels (e.g., tidal creek habitats which serve as first repositories and impact zones for terrestrial runoff and adjacent land use, Box 29).

Box 27: CDC's Harmful Algal Bloom- Related Illness Surveillance System (HABISS)

HABISS is a web-based modular system to collect and store data on human health, animal health, and environmental data in one place. This system combines environmental data with human and animal disease surveillance information and provides the basis for developing powerful new tools for the prediction, control, and prevention of adverse health effects associated with the ocean.

Collection of environmental data is combined with other epidemiologic data sets in HABISS and made available for more comprehensive analyses of human health threats. Photo credit: L. Backer.

Specific organisms or habitats may be excellent sentinels for particular questions whereas others may be sentinels for broader surveillance purposes. In most cases, they not only indicate the presence of a threat but can provide information about the possible risks or effects in the ocean environment. Ocean sentinels may provide a wealth of information to managers in addition to the mere presence of the organisms or chemicals of concern.

While sentinel species and habitats provide many advantages over routine water and sediment sampling, they also have important limitations. For example, migratory species such as some marine mammals may integrate environmental conditions over a large area, but may not provide information about specific localities or point sources of exposure. Sedentary species such as oysters and specific local habitats like tidal creeks may give a very accurate picture of localized effects of contaminants, toxins, and pathogens, but gaining a picture of more widespread impacts may require sampling of similar point source sentinels in numerous locations. While these characteristics of specific sentinel systems may appear to be drawbacks, they actually demonstrate the great diversity of animals and habitats that can potentially serve as sentinels for environmental threats. The appropriate choice of a sentinel will allow us to sample at several trophic levels and scales to develop better understanding of how contaminants, toxins, and pathogens may be transferred, bioaccumulated, and/or biomagnified through food chains that may include humans at the top. Because they integrate the broad range of environmental conditions to which they are exposed, sentinel species and habitats may give a much better picture of the cumulative health effects of degraded coastal ecosystems than any other measure. They also provide unique opportunities to study how disease-causing materials may be transferred directly from organism to organism and from mother to offspring, and the kinds of impacts associated with such transfer, including cellular and gene-level responses and disease pathogenesis. Such studies may be costly to conduct in animal models in the laboratory and are likely impossible to conduct in people. Thus, use of a broad array of sentinels may be a highly effective means of unraveling ocean health threats to humans.

Marine mammals such as dolphins, manatees, sea lions and otters provide important clues as to how the health of ocean ecosystems may affect human health. Photo credit: With permission from TOS.

Numerous marine organisms already serve as informative animal models for investigations related to human physiology and mechanisms of disease processes (see Boxes 30, 31). Perhaps best known are the squids and sea hares, studies of which have generated much of our basic knowledge about how the human nervous system works, and sea urchins, which are outstanding models for the study of biological development. A variety of fish species are used to investigate genetic influences on development and disease processes, kidney function, cancers, and many other health-related issues, and there are many other species of both vertebrates and invertebrates that provide crucial information related to human health. Marine organisms also may serve as models for understanding human diseases associated with marine-borne pathogens or biotoxins and may suggest new treatments or products. Because life in the sea is so diverse, it is certain that we will find many other species with characteristics that make them excellent subjects for observation and experimentation. The end result will be a better understanding of disease processes that affect humans and ways to alleviate human pain and suffering or reduce disease in people.

Significant opportunities for advancement using sentinel species and habitats, and marine biological models include the following:

- Use surveillance systems to collect, analyze, and interpret information on exposures to and effects of pathogens, toxins, and chemical contaminants in priority sentinel species in specific regions, such as marine mammals, bivalve mollusks, sea birds, selected fishes, and perhaps others (see Epidemiology section).
- Integrate information gained from studies of sentinel species and habitats with data from environmental observations and human epidemiologic studies to understand, predict, mitigate, and prevent ocean-borne health risks to humans, including both well-known and emerging threats and cumulative impacts of multiple stressors (see Epidemiology and Linking to Ocean Observing Systems sections).
- Optimize existing marine-animal models and identify and develop new organismal models for investigations that lead to improved understanding of human disease processes and health effects.
- Use findings from sentinel species such as marine mammals to design targeted laboratory experiments that will elucidate health risks to humans and ways to mitigate or avoid such risks.
- Use genomic and proteomic tools, including microarrays, for (a) rapid detection of multiple species of pathogens and HABs and identification and quantification of toxins; (b) integration into surveillance systems for sentinel species and habitats; and (c) research to unravel mechanisms of ocean-related health effects in sentinel species (e.g., marine mammals) that are particularly indicative of likely human effects (see also Genomics section).
- Investigate roles of marine mammals, seabirds, and perhaps other species as dispersal mechanisms for enteric organisms or antibiotic-resistant bacteria in the marine environment and as vectors of known and emerging disease threats to humans.
- Conduct research and monitoring to understand the sources, fates, effects, and human health threats of contaminants, pathogens, and toxins in sentinel habitats, including their possible roles as reservoirs of such threats; effects on ecosystem productivity, function, and resilience; and the potential use of sentinel habitats in ocean observing systems (See Linking to Ocean Observing Systems section).

Monitoring the health of sentinel habitats such as tidal creeks. Photo credit: S. Lovelace.

Box 28: California Sea Lions as Sentinels for HAB Exposure

In 1998, the strandings of sea lions along the California coast led to an intense investigation of the roles HABs play in marine mammal strandings. That event and numerous others demonstrated that domoic acid, a biotoxin produced by diatoms, can have acutely lethal and sublethal effects on marine organisms and can cause serious illness and death in humans. Some California sea lions that survived exposure to domoic acid have suffered seizures even long after the toxin was no longer detectable in their bodies. Some have also been shown to have permanent brain damage, abnormal behaviors, cardiac lesions, and reproductive problems (premature parturition, abortion, neonatal death, poor neonatal survival, and high levels of toxins in amniotic fluid). Through studies of the California sea lions, we are learning much about mechanisms of action, pathogenesis, and individual and population-level effects of domoic acid on organisms. This information will lead to better understanding of human risks and health effects that may result from exposure to domoic acid. In addition, studying what the dead marine mammals ate provided important information on the transfer of domoic acid through the food web (e.g., from sardines to mole crabs to sea lions). Understanding exposure pathways increases our knowledge of potential risks to animals and humans. Since 1998, there have been numerous other stranding/mortality events of marine mammals and sea birds caused by domoic acid along the California coast and elsewhere in the United States. Information gained from sentinel species such as the California sea lion allowed us to better understand mechanisms of action, low dose chronic exposure effects, and food web pathways and to initiate appropriate actions to protect people from this ocean hazard.

California sea lions and people share the same coastal waters and both eat at the top of the food web. Understanding exposure pathways and responses in marine mammals increases our knowledge of similar human health risks. Photo credit: T. Rowles.

Box 29: Tidal Creeks as Sentinel Habitats

Work by OHH researchers at NOAA's HML suggests that small tidal creeks may provide early warning of impaired ecosystem conditions and risks to public health and welfare. Tidal creeks are the primary connection between the land and estuaries and the first recipient of much of the runoff from adjacent highlands. These areas exhibit impairment years in advance of deeper open estuarine habitats and provide early warning of ecological and public health threats. HML scientists have found clear relationships between the level of watershed development, the ecological integrity of the creeks, and exposure risks to humans, including risk of exposure to potentially pathogenic bacteria. This research is being expanded to include sites in National Estuarine Research Reserves throughout the Southeast.

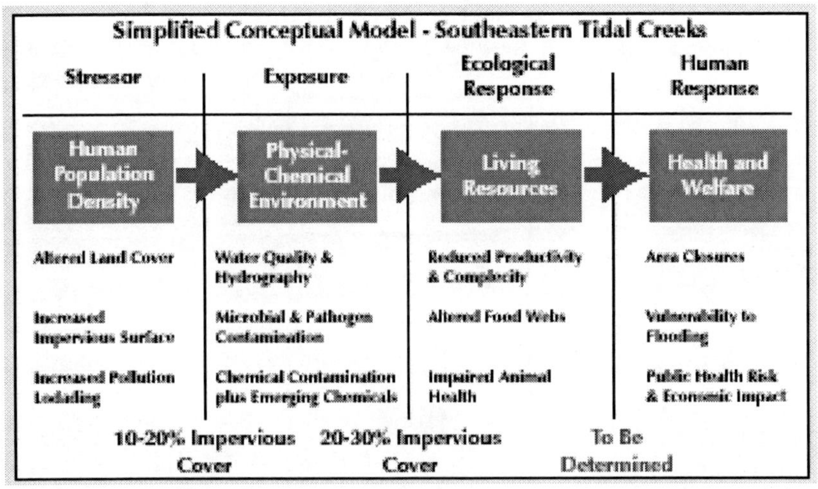

Conceptual model of linkages between altered land cover and the condition of tidal creek ecosystems based on a preliminary synthesis of previous work. Photo credit: F. Holland.

Genomics and Other "Omics" Technologies

The field of genomics is the study of organisms' genes and how they function. Genomics has revolutionized biology, leading to broad advances in biomedicine and many other fields and to the development of countless new commercial products. Not surprisingly, genomics technology is fundamental to OHH research. All seven of the OHH Centers have genomics facilities as core elements of their programs, and genomic tools are essential components of many of the research grants supported by OHH programs. In fact, many if not most of the new approaches to development of sensors for detecting pathogens and biotoxins use genomic-enabled technologies. In addition to their value in identifying pathogenic organisms, molecular tools can even detect genetic changes in pathogens that may make them more or less virulent and thus more or less likely to be a problem to humans (see Boxes 32, 33). Other related "omics" sciences, such as proteomics (the study of what proteins are produced or not as a result of gene activity and the roles of those proteins in life functions) and metabalomics (the metabolic consequences of an organism's physiological and biochemical processes), are also important for OHH research.

BOX 30: ANIMAL MODELS IN TOXICOLOGY AND MECHANISM OF ACTION - CHARACTERIZATION OF DOMOIC ACID

Researchers at the NSF-NIEHS Center for OHH at the University of Washington are interested in characterizing the molecular and cellular processes that occur following exposure to domoic acid. Animal models are used in the evaluation of domoic acid and its effects on brain developmental processes in both prenatal and postnatal phases. Specifcally, two different in-vitro cell culture systems have been established: primary mouse neuronal precursor cells (from prenatal midbrain tissues) and a cerebellar granule neuronal cell culture (from post natal brain tissues). Use of these cell cultures has allowed for the examination of critical cellular process and responses, including proliferation, differentiation, oxidative stress, cytotoxicity, and apoptosis. Characterizing the mechanism of action through animal models will assist in the identification of susceptiblity factors that can potentially define "at risk" populations, and help inform public health protection activities.

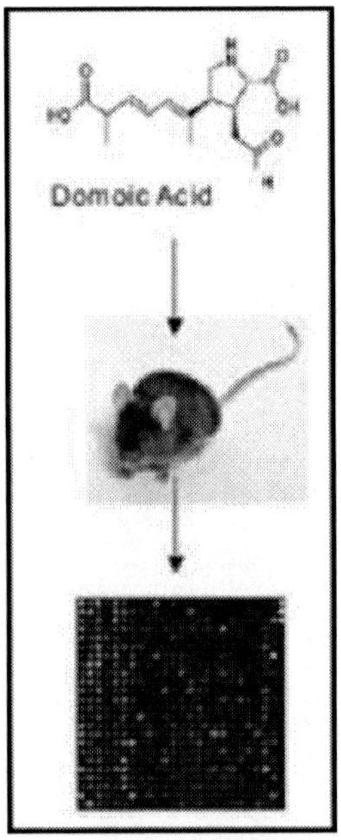

Microarrays are used to measure changes in gene expression, as shown here in an animal model using mice to help characterize molecular and cellular responses to domoic acid exposure. Photo credit: M. Vredevoogd.

In response to the needs of researchers, public health officials and environmental managers to screen aquatic environments for the presence of many different harmful algal species rapidly and simultaneously, OHH scientists at the HML have created a new technology, SIVCA (see Box 7). The new technique amplifies small, fluorescently labeled fragments of DNA from each of the target organisms, enabling their detection even at extremely low concentrations. The quick realization of SIVCA is a testament to the ready collaboration across disciplines within the OHH program.

Box 31: Use of Aquatic Biomedical Models to Assess Human Health Risk

Scientists at NOAA's WCCOHH refined an aquatic biomedical model using zebrafish to evaluate effects of polycyclic aromatic hydrocarbons (PAHs) on heart function. PAHs are chemical contaminants derived from fossil fuels, vehicle exhaust, and tobacco smoke and are highly abundant in air pollution and ubiquitous in the marine environment. Largely as a result of the Exxon Valdez oil spill, studies on the health impacts of PAHs in fish were undertaken. Research documented malformations in fish embryos, and reduced growth rates and survival to adulthood, even at very low exposure concentrations. Humans are exposed (via air pollution) to PAH mixtures with compositions very similar to those found in weathered Exxon Valdez oil. The OHH researchers used the zebrafish model to test toxicity of PAHs and found that these pollutants had significant effects on the heart, disrupting cardiac rhythm and contractility. Based on similarities between zebrafish and human hearts, the researchers suggest that PAHs may contribute to the increased heart disease-related mortality/morbidity in humans associated with air pollution. This work also demonstrates how fish can be used as both sentinel species to indicate ecosystem level and human effects of PAHs, and as biomedical models.

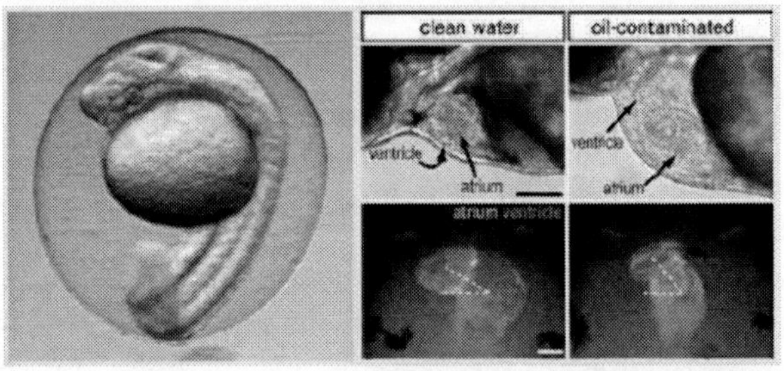

Using a zebra fish human health model, researchers found that PAHs disrupted cardiac function. Normal and abnormal heart development in larvae is shown in control embryos grown in clean water and exposed embryos grown in oil contaminated water. Photo credit: J. Incardona and N. Scholtz.

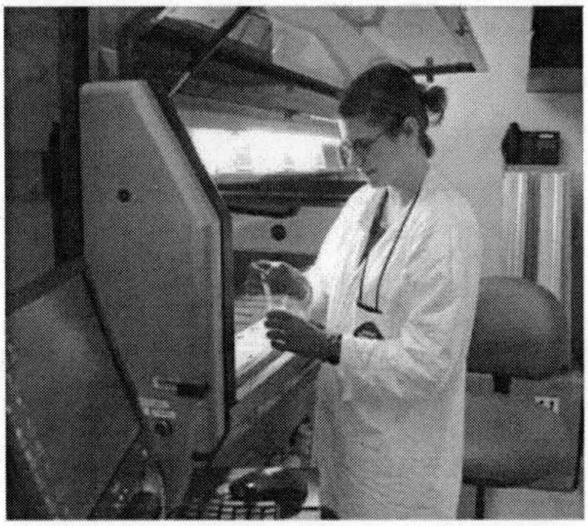

Scientists use genomics as a tool to track changes in gene function and regulation as an indicator of organism health in response to a variety of stressors. Photo credit: S. Lovelace.

Genomic tools such as microarrays, genetic bar-coding, and shotgun sequencing, provide highly sensitive means for identifying changes at cellular and molecular levels and for identifying both known and previously unknown species of pathogenic and other marine organisms. They also provide powerful tools for defining the health status of marine organisms as well as the sources of pathogens. OHH researchers are currently using such techniques to probe discrete molecular- and cellular-level responses in marine organisms that are elicited from the direct effects of one or more stressors to generate diagnostic profiles. These cellular responses can precede ecosystem-level responses by days to perhaps as much as years, providing an "eco-genomic" basis for development of early warnings and the implementation of management strategies that might prevent or mitigate the threat before it becomes a problem (see Box 34).

OHH scientists are also conducting toxicological studies aimed at identifying and understanding cellular responses to environmentally relevant exposures to marine toxins, pathogens, and chemical pollutants, including pharmaceuticals and flame retardants. These studies reveal levels of exposure (dosage) that may cause harm (response). Advances in genomics and proteomics allow researchers to investigate subtle changes that may occur at the molecular level following exposures to chronic, low levels of toxins, pathogens, or chemical contaminants. Over the long term, integration of genomic information with ecological and health effects data will enhance the sensitivity of ecosystem models and lead to a deeper understanding of the basic physiology and biochemical mechanisms of pathology for both marine animals and humans.

Genomics research produces huge amounts of data requiring robust computer and bioinformatics infrastructure for access, analysis, integration, and storage. The Moore Foundation is supporting a new program, Community Cyberinfrastructure for Advanced Marine Microbial Ecology Research and Analysis, or CAMERA, which is developing essential computational infrastructure for genomic investigations of marine microbes. Access to similar infrastructure can improve the OHH Centers' abilities to link and analyze genomic, environmental, and public health data sets and upload large genomics data sets to the National Center for Biotechnology Information where they can be accessed by other researchers. Important progress in this area has been made within the OHH program (see for example, www.marinegenomics.org), but much remains to be accomplished (see Data Management section).

Major opportunities for application of genomics to advance the broad suite of OHH research activities include the following:

- Enhance genomics core facilities in existing OHH Centers, including coordination among Centers and expanded collaboration with OHH researchers outside the Centers and develop improved cyber infrastructure and association with bioinformatics experts (see Data Management section).
- Develop or improve, demonstrate, and validate genomics-enabled sensor technologies that rapidly, accurately, reliably, and inexpensively identify and quantify bacteria, viruses, and HAB organisms of public health concern in ocean and coastal waters, including their relative pathogenicity or toxicity.
- Apply comparative genomic, proteomic, and metabolomic approaches to understand effects of exposures to chemical contaminants, pathogens, and toxins at the molecular level in marine organisms and mechanisms of toxicity and disease pro-

cesses so as to gain new insights into how humans may respond to ocean health threats.

- Apply genomic tools to diagnose and improve understanding of disease processes in sentinel species and habitats and their occurrence in response to specific environmental perturbations (see Sentinel Species/Habitats section).
- Enhance gene sequencing and mapping efforts for a broad array of sentinel and model marine species to provide a foundation for understanding biological responses to environmental perturbations.
- Use genomic, proteomic, and metabalomic tools to identify new bioactive compounds with potential for commercial development and understand how they function (see Pharmaceuticals and Other Beneficial Products section).

BOX 32: NSF-NIEHS STUDY - THE GENOMICS OF FLORIDA RED TIDES

OHH researchers at the University of Miami are attempting to gain a better understanding of the physiology and ecology of the Florida red tide organism, *Karenia brevis*, using a functional genomics approach. The objective of this work is to identify the genes that affect bloom initiation, development, maturation, and toxin production. Thousands of genes were isolated from *K. brevis* cells and used to create microarrays for assessing gene expression under various environmental conditions. With this approach, the reactions of many genes can be measured separately at one time on a single glass slide. To complement the laboratory studies, a new instrument, the *BreveBuste*r, developed at the Mote Marine Laboratory, was used to identify and measure populations of *K. brev*is in natural settings. The investigators anticipate that combining data obtained from the microarrays with those collected in the field using the *BreveBuste*r will allow them to determine what genes are important for cellular processes such as the production of brevetoxin.

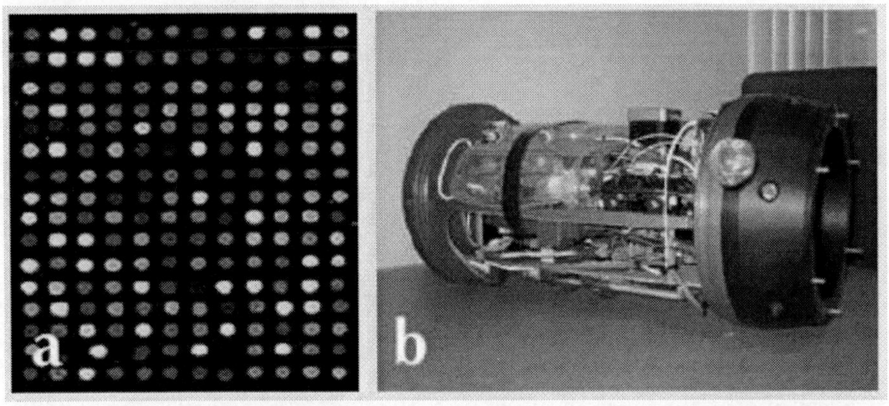

a) Microarrays contain microscopic spots of DNA samples with each spot representing a different gene. The intensity of dye in each spot is used to quantify the expression of each gene in a sample, for example in seawater. (b) The BreveBuster is a submersible device that assesses the presence of a red tide by identifying and counting *Kareina brevis.* Photo credit: a. D. Crawford, b. G. Kilpatrick.

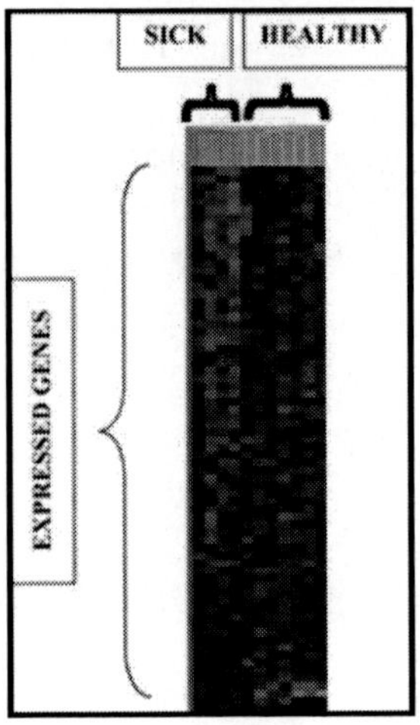

Genes are expressed differently in sick and healthy dolphins. In this expression profile, green indicates a gene is turned off and red indicates a gene is turned on. Photo credit: A. Mancia.

BOX 33: GENE-ENVIRONMENT INTERACTIONS

The NSF-NIEHS-funded PNW H2O Center in Seattle is bringing scientists from multiple disciplines together to explore interactions between genes and the environment using novel techniques. In one project, researchers are examining the effects of domoic acid on cells from mice lacking enzymes that are important in protecting cells from oxidative stress. Another study is using whole genome sequencing to identify genes that cause the alga *Pseudonitzchi*a to produce domoic acid. This project is also using expressed sequence tags (ESTs) to identify cellular pathways associated with toxin production.

Social, Behavioral, and Economics Research

The affect of the health of the ocean on humans can encompass both positive and negative economic and socio-cultural impacts including gained and lost revenue and value associated with seafood harvesting, recreation and tourism and other potential impacts to subsistence and social activities. In the U.S., in the case of HABs alone (not including freshwater outbreaks), the economic effects are conservatively estimated at $82 million per year, with public health costs of illness and commercial fisheries impacts accounting for more than 90 percent of this total. On the other hand, the potential economic value of new pharmaceuticals and other natural products from the sea could be in the billions of dollars annually.

Social, behavioral, and economic (SBE) research helps decision-makers optimize beneficial uses of coastal and ocean resources while mitigating and adapting to public health, socio-cultural, and economic impacts of stress- ors from the oceans. SBE-focused studies that relate the health of the oceans to human health are relatively sparse at all scales, from local to national. Also, little is known about the SBE effects associated with human exposures to contaminated seafood, drinking and recreational waters, and the cumulative impacts and interactions of altered environments and non-point source pollution.

Box 34: NOAA's OHHI Develops Microarray Tools to Measure Organism Health and Exposure

To quickly monitor and assess the hazards to human health posed by eating shellfish and by residential, commercial, and recreational activities in the coastal and marine environment, OHH scientists at the HML have developed new and powerful tools based on advances pioneered in the Human Genome Project. These tools, called gene chips or "microarrays," permit the simultaneous measurement of the expression of thousands of genes. Analyses of gene expression can provide a sensitive and accurate diagnosis of the health of an organism and its exposure to infectious agents or chemical contaminants that may pose risks to human health. A national and international collaboration with researchers on the Atlantic, Gulf, and Pacific coasts, as well as in Norway and France, has generated a microarray with 30,000 DNA spots to measure gene expression in oysters. This microarray tool has been made available to the research community and is currently being used in studies of oysters as sentinels of human exposure to bacterial infection and heavy metal contamination.

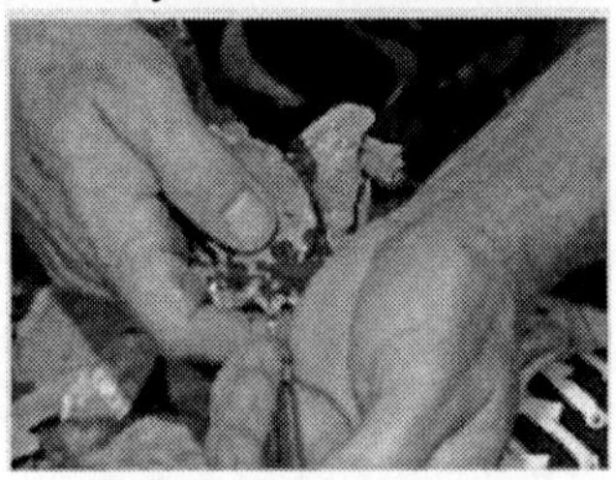

Samples are taken of oyster blood to monitor oyster health status using genomicbased tools. Photo Credit: N. Burnett.

A 2006 multi-agency report, *Harmful Algal Research and Response: A Human Dimensions Strategy (HARRHD)*, builds on the national HAB plan (*HARRNESS*) to put forth SBE research needed to improve HAB mitigation. This report provides a framework for human- dimensions research in the Interagency OHH Program as well. Opportunities to advance SBE research related to oceans and human health should focus on work in three major areas: (1) economic and sociocultural values and impacts; (2) mitigation and adaptation; and (3) integrated decision-support tools that improve policy actions made by decision-makers. Further actions include the following:

- Building on the National Ocean Economics Program and other related efforts, estimate the net economic value associated with public health benefits that humans derive from the ocean, including the economic potential related to the discovery of biomedical compounds and other products.
- Assess, monitor, and forecast economic impacts (including both market and non-market valuation) and socio-cultural impacts (e.g., disruption to cultural subsistence practices, loss of recreational opportunities, and disruption to family and community relationships) related to health threats in coastal waters.
- Estimate total economic costs for illnesses related to ocean exposures, including out-of-pocket expenses due to infection and willingness to pay to avoid being exposed to diseases, contaminants, or toxins.
- Estimate the marginal benefits and costs for specific policy options designed to prevent, control, or mitigate ocean risks to human health, and determine the value of information provided that helped lead to successful actions.
- Assess, monitor, and forecast economic and sociocultural consequences of resource management decisions designed to improve human health and welfare.
- Evaluate actual utility of OHH information to target audiences (e.g., assess how the public and officials use and value information provided from OHH research).
- Evaluate and improve the efficacy of risk-communication approaches (e.g., beach or fishery closures, seafood advisories, and public science forums) for promoting human behaviors that mitigate ocean impacts on human health and welfare (see Outreach and Education section).
- Identify human populations most susceptible to ocean-related health threats, with a focus on the importance of physiological traits, behavioral factors, occupation, economic status, and socio-cultural status, in influencing exposure and susceptibility, and assess the vulnerability of affected populations.
- Develop a suite of integrated epidemiologic/economic models that include social, behavioral, economic, and health impacts associated with human exposure to pathogens, toxins, and chemical contaminants from ocean and coastal waters (see Pathology, Contaminants, HABs, and Epidemiology sections).
- Conduct needs assessments and collaborate with resource managers, public health professionals, public service groups, and other users in developing and transitioning research applications (see Outreach and Education section).
- Develop information products and other tools that assist policy- and decision-makers in evaluating risks and benefits of various actions, including measures of uncertainty associated with the expected risks and benefits.

Climate Change and OHH

There is strong scientific consensus that Earth's climate is changing and that these changes are affecting both human and natural systems. The Interagency OHH Program can contribute to increased understanding of climate effects on ocean-related threats to human health. Among other things, shifts in ocean currents and changes in ocean temperature and global precipitation patterns may affect the distribution and prevalence of marine microbes, human and animal diseases and disease vectors, HABs and their biotoxins, and food from the sea. The effects of climate changes on distribution and bioaccumulation of chemical contaminants in marine food webs are also poorly understood and may be significant for vulnerable populations of both humans and animals. Thus, climate change is a cross-cutting issue for ocean health risks.

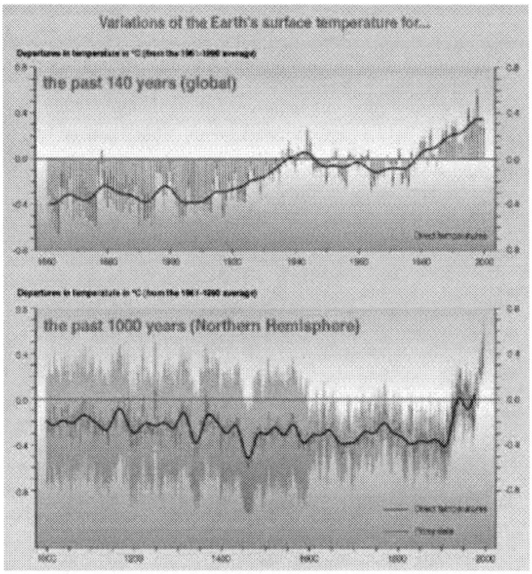

Over the last 60 years the Earth's surface temperatures have been higher than any other time in the last 1,000 years. Increases in sea surface temperature can influence the proliferation, spread and occurrence of diseases and HABs. Source: http://www.ipcc.ch/present/graphics/2001syr/ large/05.16.jpg.

Ongoing OHH activities within the NSF-NIEHS and NOAA Centers and elsewhere are well positioned to contribute to understanding how climate change will affect oceans and human health. Two existing OHH Centers are focused on OHH concerns in tropical and subtropical systems, while the geographic location of others allows a focus on temperate to northern systems. It is likely that climate change will result in warmer temperature and expansion of subtropical climates into temperate regions and some temperate expansion into northern waters. Thus, current research targeting HABs, pathogens, and other ocean health risks will contribute to an understanding of the broader effects of climate change on ocean ecosystems and health risks. In addition, ocean observing systems that include collection of OHHrelated information (see "Linking to Ocean Observing Systems"), along with robust database development (see "Improvements in Data Management and Access"), will provide the time series data helpful for addressing climate issues within an OHH context.

Specific opportunities for the Interagency OHH Research Program to contribute to advances in understanding effects of global climate change, based on existing programs and utilizing ocean observing systems and data management tools, include the following:

- Determine how the prevalence and virulence of pathogens, HABs, and contaminants are linked to ecosystem variables, including trends in global climate change.

- Include climate change information in epidemiological and disease surveillance studies for known and emerging ocean-borne human health threats.
- Consider the roles of climate cycles and change in the uptake and accumulation of pathogens, toxins, and contaminants in seafood.
- Integrate climate change data in development of models and other predictive tools.

Infrastructure

Linking to Ocean Observing Systems

EPA, NASA, NOAA, NSF, USGS, the U.S. Navy, and other federal agencies are investing significant resources in the development of national-level ocean environmental observatories. In some cases (e.g., NSF) the priority is for research, while other agencies' priorities include a variety of missions in addition to research. Currently, there are more than 40 coastal observing systems covering much of the U.S. coastline. They are operated independently or jointly by various federal, academic, state, and industry entities, and they encompass a wide range of observing and monitoring technologies, including fixed and free-floating buoys, cabled underwater stations with sensors, autonomous underwater vehicles (AUVs), remotely operated vehicles (ROVs), remote-sensing technologies, satellites, and various kinds of sensors mounted on vessels, satellites, and aircraft (see Boxes 35-38).

To provide data at the temporal and spatial scales needed, many of the observing systems are being organized into the U.S. Integrated Ocean Observing System (IOOS) that will be managed through a number of Regional Associations (RAs). These RAs are working together under a federation concept, with federal oversight to further the seven societal goals adopted by the IOOS. The IOOS goals are: (1) improve predictions of climate change and weather and their effects on coastal communities and the nation; (2) improve the safety of maritime operations; (3) mitigate the effects of natural hazards more effectively; (4) improve national and homeland security; (5) reduce public health risks; (6) protect and restore healthy coastal ecosystems more effectively; and (7) enable sustained use of ocean and coastal resources. The need to acquire OHH-related observations as an integral part of ocean observing systems was underscored by the USCOP. Possible mechanisms by which the IOOS could address its goal of reducing public health risks were the subject of an interagency workshop, *Public Health Risks: Coastal Observations for Decision Making and Management*, held in January 2006. This workshop concluded that the IOOS could make vital contributions to public health management and play a major role in developing opportunities for protecting human health by linking OHH activities and programs with ocean observing systems.

The inclusion of robust nearshore and inshore observing networks that can provide data of immediate utility to people living, working and recreating along our coasts will ensure that IOOS is relevant to public health concerns. Automated monitoring of coastal environments could facilitate the examination of broad-scale patterns of pathogenic organism distributions, and could provide real-time or near-real-time data at spatial and temporal scales that are not possible with current monitoring methods. Biological sensors that could be located on buoys, AUVs, ROVs, and other platforms (including satellites) to gather accurate and timely data on pathogens, chemical contaminants, and HABs are essential elements for an operational IOOS. The Interagency OHH Program includes excellent work on the development of a variety of biological sensors. However, much remains to be done, and there are many opportunities to further develop these and other technologies for eventual deployment as part of regional and national aquatic observing systems.

Although some biological sensors exist, most of the specific interest for OHH research are still under development, in contrast to the increasing availability and sophistication of instruments for obtaining physical oceanographic and atmospheric measurements. A variety of these (e.g., acoustic Doppler current profilers, surface current radar, satellite-tracked surface drifters, submerged drifters, moored thermistors, and conductivity probes), as well as a range of other oceanographic and meteorological sensors, will be included in the IOOS and will provide essential information for interpreting biological observations. In addition, hydrological data (e.g., rainfall, runoff, streamflows, water temperature and color, currents, soils) provide essential input for hydrographic models that may be employed to forecast transport and distribution of pathogens, chemical contaminants, HABs, and toxins in estuarine, coastal, and Great Lakes environments. Implementation of the National Water Quality Monitoring Network, as called for in the *Ocean Action Plan*, and its linkage to the IOOS will improve estimates of land-based inputs of water, sediments, nutrients, and contaminants and their fates once they enter estuarine and coastal environments and will provide crucial information to enhance understanding and forecasting of ocean health threats.

Across the board, technological advances are expected to result in increasingly rich and complex data streams that will produce a broad variety of physical, chemical, and biological information. Major foci for development include enhancements to facilitate networking of disparate sensors and platforms, processes, and infrastructure for combining very different data sets and advanced data management, integration, modeling, and synthesis techniques. Data Management and Communications (DMAC) is one of the highest-priority areas of the IOOS (see Improvements in Data Management and Access section).

Federal agencies are also planning, operating, or participating in other large-scale, coastal and open-ocean observing systems and networks such as the NSF-supported Oceans Observatories Initiative (OOI – See Box 36) and the National Ecological Observatory Network (NEON). The OOI will be operated by and for researchers and driven by basic research questions related to how the earth-ocean-atmosphere system works, while NEON is envisioned as a continental-scale system of sensor networks, experimental infrastructure, and remote sensing linked to a computational, analytical, and modeling facility. As OOI and IOOS mature, the design and output of each will benefit the other, and the technological advances expected from NEON are likely to be applicable to ocean and coastal observations important for addressing OHH-related issues.

The next generation of NASA's Earth-observing satellites will include the Ocean Surface Topography Mission (extends the collection of sea surface topography data), Aquarius (measures global sea surface salinity), and the Global Precipitation Mission (measures the variability of global precipitation to advance knowledge of the Earth's water/energy budget). These satellites will provide oceanographic and hydrographic observations of even greater temporal and spatial resolution than currently available.

Major opportunities for advancement by linking the Interagency OHH Program with ocean observing systems include the following:

- Increase participation in the IOOS regional as sociations by coastal managers, public health practitioners, and stakeholders responsible for beach and shellfish management to help optimize observing-system capabilities that provide useful human health-related information products.
- Develop and incorporate sensors for pathogens, indicator microorganisms, harmful algae, toxins, other contaminants, and water quality into the observing systems at local to regional scales to acquire essential data for public health protection.
- Develop and provide multi-sensor data products in near-real-time for public health applications (e.g., detection, fate and transport of pollutants/pathogens, and blooms) and develop new and improved information products to support beach and shellfish-bed opening and closure decisions.
- Collect data in nearshore environments where the majority of the public's interaction with coastal waters occurs; record environmental observations at locations, and temporal and spatial scales particularly relevant to the population dynamics of waterborne pathogens and harmful algae.
- Monitor migratory species, including seabirds, fish, and marine mammals, to improve assessments of the health of these species and their potential to serve as vectors for, or indicators of, threats to human health.
- Strengthen efforts to make collection of epidemiologic, public health, and animal health surveillance data integral parts of the ocean observing systems, including the archiving of both environmental and health surveillance information to enable retrospective analyses (see Epidemiology section).
- Collect physical and remotely sensed oceanographic and hydrographic data, including stream and river flow, non-point-source runoff, and their chemical constituents for estuarine and coastal areas and make these data readily available to researchers for use in model development and testing.
- Develop and provide data and information needed to quantify relationships between changes in land use and land-based inputs to coastal waters, and changes in public health risks.
- Continue development and deployment of Earth science satellites for collection of essential ocean- ographic and hydrographic observations at greater temporal and spatial resolution than currently available (see Box 37).
- Following recommendations of the international Group on Earth Observations, work with the World Meteorological Organization and other international bodies to explore how observing systems may respond to needs of the public health community by collecting, providing, and validating remotely sensed environmental measurements relevant to human health, including the possible initiation of pilot projects integrating Earth observation, health, epidemiologic, and socio-economic data.

BOX 35: IOOS AND PUBLIC HEALTH

Designed to be user-driven and provide output in forms and rates useful to decision-makers, the IOOS will link a wide variety of remotely sensed observations and other data collection methods, data management, and modeling to provide useful products on local to global scales, for example, from the local scale of beaches and shellfish beds to the global scale of an El Niño event. Remote-sensing capabilities include satellite-, aircraft- and land-based sensors, power sources, and transmitters. Direct sensing includes use of platforms (e.g., ships, buoys, gliders), in-situ sensors, power sources, sampling devices, laboratory-based measurements, and transmitters.

Ocean observing systems link data collected from multiple platforms such as ships and in-situ monitoring stations with remote sensing capabilities from satellites, submersibles and aircrafts. Source: Ocean.US.

BOX 36: A CONCEPTUAL DIAGRAM OF THE NSF OCEANS OBSERVATORY INITIATIVE (OOI)

The OOI will have three observational components: coastal observatories; a regional cabled observatory; a global system of relocateable moorings for open ocean settings at high and low latitudes. Cyberinfrastructure will link OOI to research and education applications. OOI will provide infrastructure to enable hypothesis-driven basic oceanographic and geophysical research by fostering specialized observations, instruments and activities for the purpose of answering basic research questions. The high-speed, interactive communication capabilities of OOI will open entirely new avenues for student and public audiences to interact with and understand the ocean.

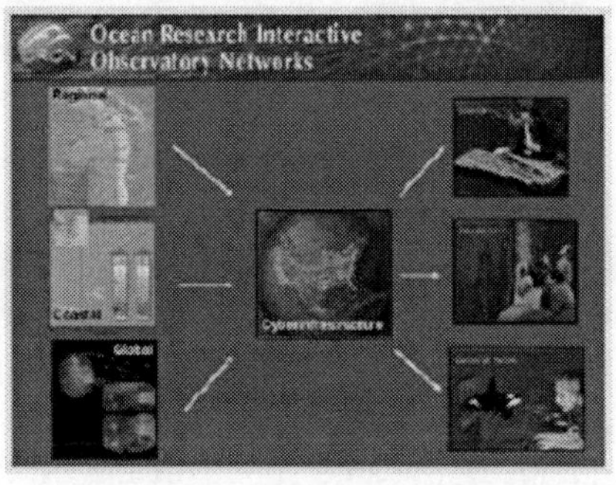

Source: NSF

Box 37: NASA Earth Observing Research Satellites – Critical Components of Ocean Observing Systems

NASA's Earth Science Division works to understand global, integrated Earth system science, including the response of the oceans to natural variations and human-induced changes. To this end, NASA developed and deployed a constellation of Earth-observing research satellites named Aqua, Aura, Terra (provides data on ocean properties including ocean chlorophyll and sea-ice cover) GRACE, Jason, QuikScat, TRMM, and others, along with aircraft and surface-based sensors. Essential oceanographic observations are collected from satellites such as Jason (collects sea surface topography data), GRACE (acquires data on variations in Earth's gravity field due to ocean currents and exchanges between ice sheets/glaciers and the oceans; its measurements yield vertically integrated ocean circulation), and the Moderate Resolution Imaging Spectroradiometer (MODIS) sensor on Terra and Aqua. NASA also created the world's largest data and information system for collection, processing, archiving, and distributing Earth-observing satellite data.

Improvements in Data Management and Access

OHH is a broad interdisciplinary field not limited to databases within a single scientific domain, such as estuarine ecology or animal physiology, but instead encompasses aspects of Oceanography, ecology, genomics, human health, and numerous other fields. Each of these disciplines has advanced, often program specific databases, oriented toward their specific needs. These data archives may be available online but are not necessarily integrated with or linked to other program-specific databases. Similarly, public health databases relevant to the OHH community reflect the breadth of public health science from the molecular level to individual and population studies and cover a broad range of health effects. The capability to access and link such data from disparate agencies and programs is critical for retrospective analyses and determining trends in ecosystem condition and human health.

Massive volumes of data are already being collected from observing platforms, satellites, genomics investigations, various public health studies, and these streams will continue to expand exponentially. Because data management is so central to the entire ocean observing system, the IOOS DMAC community is developing Web-based networking capabilities to meet the IOOSspecific needs. The *Guide for IOOS Data Providers* presents a draft of the IOOS Data Policy (available on the Ocean. U.S. Web site), along with the most recent (2006) guidelines for IOOS data and metadata interoperability standards and best practices. These will be a valuable resource for the OHH since the developing OHH data structure which should link seamlessly with the IOOS (see Linking to Ocean Observing Systems section and Box 39).

Although progress is being made on the broad scale, the OHH community must establish a common "language" for its database requirements and determine what data need to be accessed, as well as the requisite "metadata" (i.e., the who, what, where, when, why, and how of the data itself). The Federal Geographic Data Committee (FGDC) Content Standard for Digital Geospatial Metadata (CSDM) is widely used for geospatial data, and should be applicable to some OHH data. According to the DMAC, the IOOS will initially use the FGDC Content Standard, and any applicable supplemental profiles (i.e., the Biological Data Profile, Shoreline Profile), as its standard for metadata. However, the OHH research community may have to invest considerable effort to develop procedures to crosswalk any other data standards in use with the FGDC and CSDM standards.

Sharing data across disciplines is fundamental to advancements in OHH research and predictions. Data must be capable of being easily accessed and shared electronically. These requirements suggest a data-sharing system that engages a diverse spectrum of data providers and users. A number of integrating technologies show potential for use with OHH data. For example, a multi-federal partner program at Oak Ridge National Laboratory is leading the development of SensorNet, a framework that ties together sensor data from all over the country to create a real-time detection and alert system for various threats, whether chemical, radiological, biological, nuclear, or explosive. Although the SensorNet effort is primarily focused on homeland security, the technology could be adapted to integrate data from estuarine, coastal, and ocean monitoring systems and other data streams in the Interagency OHH Program. Similarly, "grid" technology allows users to share distributed and heterogeneous computing resources in many research areas from remotely sensed data and climate modeling to bioinformatics and analyses of human disease. In addition, metadata and disparate databases need to be integrated in a manner that allows for automation. The developing Semantic Web provides a common framework allowing data to be shared and reused across application, enterprise, and community boundaries.

Box 38: NASA MODIS

The Terra and Aqua Earth observing satellites acquire data on ocean chlorophyll fluorescence. Anomalies in ocean chlorophyll fluorescence are used in the detection of HABs.

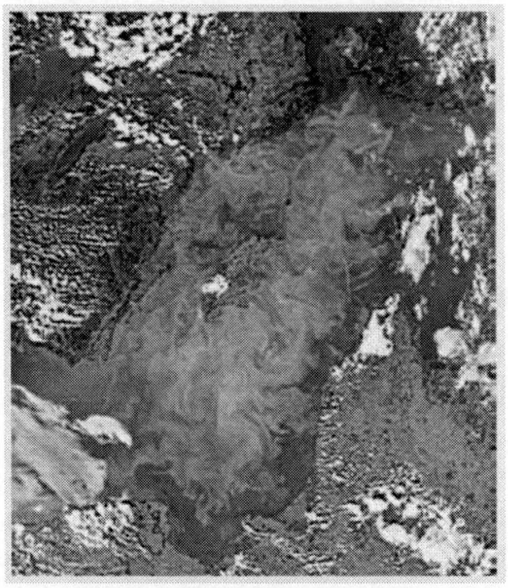

This MODIS image from July 8, 2005, shows a potential harmful bloom in the Baltic Sea. Photo credit: NASA.

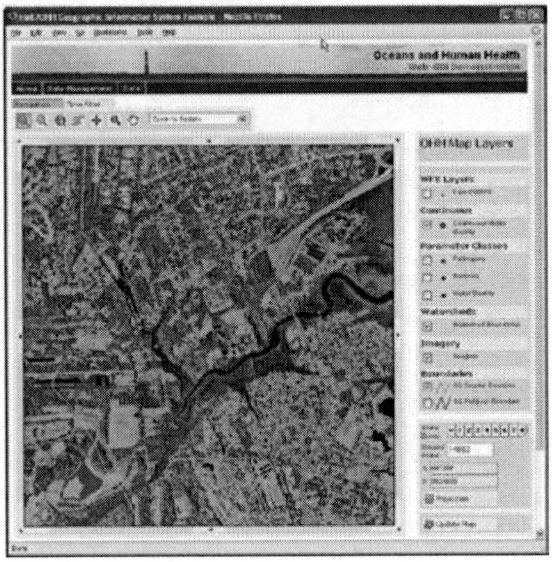

OHH data are assimilated into common databases where multiple inputs can be layered, such as in these Geographic Information Systems (GIS) maps which are used to help managers make informed decisions. Photo credit: D. White.

Acquiring and assimilating the necessary data into products useful to decision-makers, and increasing the knowledge base of the public at-large with regard to the connection between ocean health and human health are critical. The basic technical challenge is how to best apply modern information technology to effectively link numerous, distributed databases of variously formatted data. This effort will include determining minimal standards for long-term data archiving and implementing ways by which users can readily discover data, determine their availability, and easily gain access and use the data. An additional challenge is to foster understanding among the interested communities that the limitations of resources likely to be available to data providers will restrict the ability to make all data equally accessible and linkable all of the time.

Major opportunities to enhance OHH data management include the following:

- Conduct interagency and community workshops to identify databases with high relevance to OHH research and their structures/content across disciplines, perform a baseline data needs assessment, define overall OHH database requirements, and provide recommendations on how best to create the OHH architecture to complement other ongoing data management efforts (e.g., under IOOS).
- Engage the public health community in OHH data management planning. This dialog will enable development of standards and processes, allowing data important for public health to be accessed, transported, and integrated into models and products for risk assessment that are useful to public health officials and the general public (see Outreach and Education section).
- Develop and implement essential cyber infrastructure to enable data sharing, integration, analysis, and access, with a particular focus on processes for linking and sharing environmental, genomic, and epidemiological data.

- Coordinate data management and access activities with the IOOS DMAC to ensure compatibility (see Linking to Ocean Observing Systems section).
- Develop and validate coupled physical-pathogen transport models for nowcasting risks and forecasting changes in risks with known accuracy.
- Develop and improve nearshore circulation models that link land-based inputs and nearshore processes with better offshore boundary condition information. Incorporate pathogen and algal biology into these models.

BOX 39: NOAA'S OHHI DEVELOPS MODELS TO INCORPORATE NUMEROUS DATA LAYERS

NOAA's CEGLHH is developing a refined model to incorporate numerous data layers to predict the hydrodynamics of the Great Lakes and assist public health forecasting. In the Great Lakes, one of the biggest issues relating to human health is exposure of bathers and other beach users to bacteria stemming from creeks and rivers that convey surface runoff and combined sewer overflow discharges. Relevant data layers include information related to hydrodynamic processes (e.g., wind speed and direction, water temperature, current speed and direction, topography, biological contaminants, and chemical contaminants). These data layers are being combined in a GIS-based model to predict bacterial concentrations on recreational beaches.

Standards and Standardized Methods

Establishing data validity is the driver behind the development and use of standards for physical, chemical, and biological measurements as well as for data management and bioinformatics (see Improvements in Data Management and Access section). Standards are an integral component of monitoring programs important to protecting human and environmental health. For example, EPA's water quality standards define the goals for a water body by designating its uses, setting criteria to protect those uses, and establishing provisions to protect water quality from pollutants.

Methods and procedures must be validated and standardized, particularly when they are to be used in regulatory or management contexts. Professional associations, such as the AOAC International, usually validate analytical procedures. AOAC has a Marine and Freshwater Toxins Task Force devoted to accelerated validation of methods for analysis of HAB toxins. Similarly, the ISSC has established a Laboratory Methods Review Committee to help validate methods for ensuring the safety of shellfish intended for human consumption.

Another aspect of establishing credible measurements involves establishing procedures to produce comparable data. The National Institute of Standards and Technology (NIST) coordinates a number of performance-based quality assurance (QA) programs in micro-nutrient measurement, marine monitoring, marine mammal analyses, and other areas. These programs have resulted in dramatic improvement in quality and comparability of data. Although often overlooked, other tools historically used for QA are archived sample and tissue banks, including NIST's Environmental Specimen Bank, CDC's sample banks from disease outbreaks, and the National Cancer Institute's bank of bioactive materials. By maintaining subsets of samples collected by ongoing research and monitoring programs, specimen banks play an increasingly important role in assuring the quality of analytical results and in providing historical materials for comparison with new samples.

New and emerging ocean health threats dictate a need to enhance and/or expand existing standards and methodologies. Although Certified Reference Materials are readily available for analysis of well-known contaminants, few standards are obtainable by researchers working with more recently recognized chemicals of concern (such as brominated flame retardants, fluorinated compounds, and metabolites of pharmaceutical compounds). Work is currently underway in this area by OHH and NIST researchers, but a great deal remains to be done before these compounds can be analyzed routinely. Another critical issue is the need for a better understanding of the quality of microarray results to support studies on gene expression. NIST, in collaboration with OHH researchers, is beginning to develop standards for the production and generation of genomic microarrays, including protocols for performance and analysis.

Targeted efforts to improve and validate measurements and standards are needed for pathogens, pathogen source tracking, HAB toxins, and pharmaceuticals. Although several methods may help identify and differentiate bacteria from human and wildlife sources, uniform standards that would make each method more proficient and reliable, especially as they are applied in different watersheds, are lacking. Similarly some brevetoxins, which are produced by the HAB species *Karenia brevis*, can be detected by an enzyme-linked immunosorbent assay (ELISA) method in biological samples; however, ELISA of biological samples is not currently a certified method for detection of brevetoxin.

Removing samples from a liquid nitrogen vapor phase freezer in the Marine Environmental Specimen Bank maintained by NIST. Photo Credit: R. Pugh.

Credibility of measurements resulting from OHH research will hinge on careful documentation of procedures used, quality assurance efforts, validation, and standardization of material and methods. There are many opportunities for advancement of OHH research in this area, including the following:

- Develop partnerships among the Interagency OHH Program, NIST, and others as appropriate to establish standards and standard protocols for emerging chemical contaminants, HAB toxins, pathogens, genomic microarrays, and ELISA procedures for OHH research and management uses.
- Develop, validate, and distribute specific standards for DNA extraction techniques (see Rapid Response Capabilities section).
- Expand performance-based QA activities across the Interagency OHH Program and implement national standards and protocols for measurements, data management, communications, and modeling.
- Improve and validate sample processing procedures to quickly and efficiently isolate and concentrate microbes from large volumes of water so as to enhance use of genomic techniques for rapid detection, quantification, and tracking of microbial pathogens (see Pathogens section).
- Develop new procedures and opportunities for long-term storage and archiving of isolates and mixed culture collections, particularly of HAB organisms and other microbes and pathogens (including viruses). Establishing centralized collections that all researchers could access would facilitate assay optimization and comparisons (see HABs section).
- Improve infrastructure needed to protect human health during HAB events and for research on HAB impacts. Make available certified toxin standards and radiolabeled toxins, protocols, and methods for toxin analysis, and increase capacity for rapid analysis of known toxins and identification of unknown ones (see HABs section).
- Expand specimen and sample banks to ensure maintenance of appropriate materials to evaluate apparent emerging health threats in the marine environment, including mechanisms to archive and provide access to bioprospecting samples (see Pharmaceuticals and Other Beneficial Products section).
- Conduct workshops to assist regulatory agencies in the development of appropriate pollution standards and indicators to protect coastal and ocean water quality and prevent threats to human health caused or exacerbated by ocean pollution.

Access to the Sea to Support OHH Sampling and Exploration

Much data can be collected efficiently through platforms such as satellites, but a great deal of OHH research also relies on direct collection from ocean sediments, water columns, sea surfaces and the water-air interface. In nearshore waters, samples for public and environmental health monitoring programs may be collected by wading, from docks, piers, and jetties, or with the use of small- to medium-sized boats. In addition to sensors placed on buoys and satellites, ROVs and AUVs increasingly provide new tools for both sampling and observing in shallow to very deep seas. However, for many studies, specialized research vessels are used, in some cases to transport, deploy and operate ROVs, AUVs, buoys, and sensors (see Linking to Ocean Observing Systems section).

The primary access for academic scientists to marine research vessels is through the University-National Oceanographic Laboratories System (UNOLS). Both agency and university researchers can also take advantage of a limited number of mission-dedicated research vessels operated by federal agencies, particularly NOAA. However, their cruise missions are generally planned at least a year in advance, and are used to support ongoing agency research and monitoring objectives, as well as OHH activities. The Interagency OHH Program depends on using assets from the UNOLS system, NOAA and perhaps other agencies, and smaller, local vessels when appropriate.

Because many OHH activities are near the shore, small (< 130 feet) and intermediate-size (130-180 feet) vessels are often the most cost-effective platforms. In some cases, depending on the specific work and location, academic and/or state/local institutional vessels can be employed, principally for collecting samples that are returned to shore-based labs for analysis. These platforms are generally not amenable to long, deep-water cruises, where large scientific parties are required to support broad cross-disciplinary projects, or to carry out studies where a considerable amount of sample processing and analysis is to be carried out onboard, including bioprospecting for useful marine products.

Most, if not all, of the present NOAA and NSF-NIEHS OHH Centers utilize some form of vessel support. Currently the NSF-NIEHS Centers at the Woods Hole Oceanographic Institution and the University of Miami currently are budgeted to support cruises on UNOLS vessels. For these two Centers, using a mix of UNOLS ships and institutional vessels has proven to be a cost- effective mechanism to address sampling needs. The other Centers have leveraged ship time on other projects and/or have received logistic support from local partners or used smaller, less-specialized vessels depending on their needs. In addition, NOAA Centers may submit ship use requests to the NOAA Fleet Council, but generally requests must be submitted one to two years in advance.

Researchers and outreach specialists communicate important OHH findings to multiple audiences including the scientific community, beach and seafood managers, and the public. Photo credit: S. Lovelace.

Opportunities to provide access to the sea for OHH researchers include the following:

- Anticipate and include ship costs as part of supported research activities in OHH research projects.

- Identify, in a timely fashion, long-term needs for vessels, other platforms, and specialized equipment to support OHH research activities.
- Anticipate platform needs for OHH researchers to enable rapid response to episodic events such as HABs and hurricanes (see Rapid Response section).

Transition of Research to Application

Outreach and Education

The USCOP *"An Ocean Blueprint for the 21st Century"*, the NSTC JSOST *"Charting the Course for Ocean Science in the U.S. for the Next Decade"*, and other recent reports have emphasized the importance of outreach and education to improve the nation's science and ocean literacy and achieve healthy oceans and people. Similarly, the OHH Act of 2004 requires the Interagency OHH Research Implementation Plan to "establish goals and priorities for Federal research which... provide usable information for the prediction of marine-related public health problems and the use of the biological potential of the oceans for development of new treatments of human diseases and a greater understanding of human biology." The Act also places specific responsibility for information dissemination and outreach on NOAA.

Carefully targeted and coordinated outreach and education efforts can result in three critical outcomes for the nation:

1) Use of oceans and human health information and products by resource and public health managers to make decisions that help ensure healthy and productive marine ecosystems and people (see Box 40);
2) A dynamic, diverse, and interdisciplinary workforce with competencies critical to advancing oceans and human health research, both now and in the future (see Boxes 41 and 42); and
3) A well-informed public that understands the effects of the oceans on human health and acts as a steward of coastal and marine ecosystems, thereby preventing and reducing human health risks and maximizing human health benefits (see Box 43).

It is critical that resource and public health managers help guide OHH research needs and in turn be able to understand, access, and use information, tools, technologies, and products produced by the OHH research community to make informed decisions. Strengthening partnerships and dialogue between these managers and the OHH research community will enable effective development, transfer, and use of OHH information.

Development of a new generation of interdisciplinary scientists who are comfortable working at the nexus of ocean sciences and health sciences and able to integrate data and approaches across fields is also critical. This new cadre of undergraduate, graduate, and post-doctoral students must be trained in effective communication and partnering with multiple user groups, as well as in science. Because people learn in a variety of places from formal academic settings (such as schools) to informal locations (such as aquariums, zoos, and even their own homes), OHH education materials, activities, and practitioners must be equally diverse. Expanded efforts with the media can also increase visibility and public awareness of linkages of human health to ocean stewardship. Accomplishing the OHH outreach and education outcomes will require a dynamic, performance-based program, with effective evaluation and assessment mechanisms.

The OHH programs in NOAA and NSF-NIEHS have initiated outreach and education activities that communicate and deliver OHH information, tools, and products to a broad range of users, the media, and the public; assess information and product needs of managers and educators; and provide diverse training opportunities for students, teachers, and scientists. Other agencies involved in OHH issues, such as the CDC, EPA, FDA, NASA, MMC, and USGS conduct various outreach and education activities to inform managers and the public about ocean-related human health risks.

Major opportunities to advance outreach and education for the Interagency OHH Program include the following:

- Ensure coordination of OHH-related outreach and education activities within and among agencies in the Interagency OHH Program and leverage existing outreach-education partnerships such as the Centers for Ocean Sciences Education Excellence (COSEE), the National Sea Grant College Program, the National Oceanographic Partnership Program (NOPP), and the Interagency Working Group on Ocean Education.
- Effectively communicate with, respond to needs of, and provide appropriate OHH information, tools, technology, products, and training to resource and public health managers and other users to support improved public health and coastal and marine resource decision-making.
- Establish, maintain, support, and enhance partnerships among governments and management bodies; academia; nongovernmental organizations; professional societies; informal outreach and education facilities, including aquaria, museums, and zoos; and the private sector to communicate OHH information and products to diverse audiences and to derive information from monitoring efforts involving the public.
- Sustain and enhance the development of undergraduate, graduate, and post-doctoral students pursuing OHH-based careers; encourage scientists in the early stages of their careers to pursue interdisciplinary OHH research by supporting fellowships, traineeships, and other opportunities, such as interagency exchange programs for researchers and managers.
- Expose a wider cross section of students to interdisciplinary OHH ocean sciences and inspire youth to pursue ocean-related careers by providing professional development opportunities to educators, developing more avenues for underrepresented and underserved students, and promoting partnerships and resources to support OHH curriculum development.
- Provide information for the preparation of updated clinical medical guides and training that describe ocean-derived illnesses in humans and sentinel species (including pathology, diagnostic methods, and available treatment options) and make them widely available to public health officials, primary health care providers, veterinarians, and emergency response personnel (see sections on Pathogens, HABs, Epidemiology, and Chemical Contaminants).

Box 40: Public Health and Florida Red Tide - From Remote Sensing to Poison Information

In a unique collaboration, the Florida Department of Health (FDOH), CDC, and its public and private partners have established a linked network of public health information coupled with exposure and disease surveillance on Florida red tides. NOAA (coastwatch.noaa.gov/hab/ bulletins _ns.htm) and Florida Wildlife Research Institute (866-300-9399) produce weekly reports of the presence and location of Florida red tide blooms based on remote sensing and water monitoring that are available by phone and on the Web, while Mote Marine Laboratory (MML) has developed real-time beach condition reporting (http:// coolgate.mote.org/redtide/). This exposure information is linked to South Florida Poison Information Center Hotline (888-232-8635), which provides 24 hour toll free health information in multiple languages with funding and backup by the CDC, FLDH, and a NSF-N IEHS funded Center at the University of Miami. The University of Miami, CDC, FDOH, MML, and a nongovernmental organization, START (http://www.start1 .com), have developed beach signage, museum displays, information cards, healthcare provider information, public health recommendations and response plans, and a traveling exhibit to communicate up-to-date information to Florida's tourists and residents.

Web sites such as Solutions to Avoid Red Tide (START) are important tools to increase public awareness of ocean health risks and build support for increased monitoring, regulation and prevention of health threats.

BOX 41: TEACHERS AT SEA AND IN THE LAB

To improve integration of OHH topics in high school curricula, NOAA's OHHI Centers partnered with the Armada Project and COSEE to host educators on research cruises and in laboratories. Educators worked side-by-side with scientists to learn more about OHH topics and research methods.

Science teachers get the opportunity to work with OHH reseachers in the field, as shown here in the tidal creeks of SC. Photo credit: S. Lovelace.

BOX 42: STUDENT AND VOLUNTEER ENVIRONMENTAL MONITORINGEFFORTS

A number of observing and monitoring efforts, such as FDA's Monitoring Network (SEPMN), routinely use information provided by public volunteers. This approach not only helps build an ocean stewardship ethic in the public at large and involves students and the public directly in research, it also provides a wealth of useful information to better track the occurrence of health threats in the oceans. SEAPORT uses networks of volunteers to gather information that state shellfish programs can use to focus toxicity monitoring on times, locations, and toxins of greatest concern, thus making the overall program more effective and less expensive than would otherwise be possible. Similarly, SEPMN unites over 65 volunteer groups with scientists to monitor marine phytoplank-ton community and HABs in the southeast. The majority of volunteer groups include teachers and students in grades 5 through 12; however, universities, aquariums, parks and recreational facilities, and environmental and citizen groups participate as well. To date, over 40 HAB events have been identified and valuable data has been generated on species composition and distribution in southeastern coastal waters.

Students can be actively engaged in water quality monitoring, as a fun, educational, and valuable activity. Photo credit: C. Dolan.

Box 43: OHH Featured on The Weather Channel and in the LA Times

In 2005, NOAA partnered with The Weather Channel to produce an OHH special that aired nationally and highlighted research at NOAA's OHHI Centers. The Weather Channel continues to run this program occasionally, improving public awareness of OHH issues. In 2006, the LA Times produced a five-part series of feature articles on the crisis in the seas entitled, "Altered Oceans." Two of these articles were informed in part by a number of NOAA conferences related to marine mammal health and HABs, including a symposium at the 2006 American Association for the Advancement of Science (AAAS) involving OHH researchers focused on the effects of diseases and toxic algal blooms on sentinel species and coastal residents.

The LA Times series "Altered Oceans" by Kenneth R. Weiss, Usha Lee McFarling and Rick Loomis focused on several OHH themes and has won many awards including the Pulitzer for Explanatory Reporting. Source: LA Times.

Rapid Response Capabilities

For decades, state, federal, and local authorities have implemented disaster planning and responded to physical, infectious, chemical, and weather-related events. Since the terrorist attacks of September 2001, significant federal, state, and local efforts have been made to improve emergency readiness, planning, and response. Federal efforts have been developed under the National Response Plan and are continuing under the National Preparedness Goal.

Although such planning has been triggered by the threat of terrorism, the strategic goal has been to enhance ongoing response planning and capabilities for all hazard events from small local occurrences to disasters of national significance. These hazards include natural catastrophes such as hurricanes, outbreaks of marine- related diseases in humans or animals, and HABs, as well as human-caused disasters such as oil spills or hazardous chemical leaks. Numerous partners are involved in responses to these events at local, state, regional, and national levels, and a major goal has been to integrate these numerous players into a cohesive response network that can be managed (i.e., using the National Incident Management System) and expanded or contracted depending on the circumstances. The Interagency OHH Program can enhance these response capacities by providing additional scientific expertise in marine and human health-related sciences and laboratory capabilities to handle the surge of samples that often accompanies environmental disasters.

Participants in the Interagency OHH Program have engaged in both response planning and capacity building over the last three years and have been directly involved in responding to some recent high profile events, such as Hurricane Katrina and the 2005 HAB event in New England waters, specifically in Massachusetts Bay (see Boxes 44 and 45). In these situations, the Interagency OHH Program provided additional scientific support and increased laboratory analytical capacity to the primary responding federal, state, and local agencies. Federal agencies can provide unique expertise and capabilities to assist or lead activities to prepare for, respond to, or recover from an event. Often a federal emergency response is in response to a request from an affected state and may involve such entities as the CDC, EPA, FDA, FEMA, NOAA, USCG, and others. However, each agency's response is mission-dependent and their capabilities differ. Identification of capabilities and capacity for response within the research elements of the Interagency OHH Program would provide response agencies with additional resources when needed.

Based on the lessons learned from the Hurricane Katrina and HAB -event responses, opportunities for advancement include:

- Identify OHH research, analytical, diagnostic, and other response capabilities that could be brought to bear in emergencies and how they might be used to best complement and enhance established state, federal, and local agency disaster-response capacities.
- The Interagency OHH Program, working with NIST, federal response and regulatory agencies, and other entities, could lead efforts to develop and standardize techniques for specific marine analyses (e.g., DNA extraction) so as to improve the capacity for integrated interagency efforts in disaster responses (see Standards and Standardized Methods section).
- The Interagency OHH Program could help bring together key players (e.g., federal and state agency representatives, public health officials, academic researchers, constituents, and users groups) to evaluate different agency determinations of "safe levels" of various contaminants in seafood, water, sediments, and air and make recommendations for integrated and standardized risk communication after disaster events (see Seafood section).

BOX 44: INTERAGENCY RESPONSE TO ASSESS ENVIRONMENTAL IMPACTS OF HURRICANE KATRINA

In August 2005, Hurricane Katrina devastated the north-central Gulf of Mexico coast. Federal, state, and local agencies along with non-profit organizations and citizens rushed to aid the area. EPA, FDA, NASA, NOAA, NSF, and USGS initiated an integrated response to assess environmental impacts of Katrina throughout the affected region, with much of the effort directed towards characterizing the magnitude and extent of coastal contamination. Response activities were conducted using EPA's vessel *OSV Bold,* NASA's Experimental Advanced Airborne Research LiDAR sensor, NOAA's *R/V Nancy Foster*, the NSF-supported UNOLS vessel *Cape Hatteras*, and FDA small boat teams. Specific objectives included measuring chemical and microbial contaminants in water, sediments, and fish and shellfish tissues to assess ecological impacts and potential threats to human health. OHH researchers from NSF-NIEHS Centers, Louisiana State University, and NOAA responded to concerns about pathogens and indicator organisms in soil around New Orleans, Lake Ponchartrain, and nearshore waters following the dewatering of New Orleans, and completed a comprehensive study of pathogenic viruses, bacteria, and protozoa as well as fecal indicator organisms in the area. Also, the University of Washington OHH Center developed and distributed to returning residents an informational flier that included important messages about anticipated hazards from multiple sources inside and outside of their homes.

Hurricane Katrina Photo credit: NASA.

BOX 45: INTEGRATED RESPONSE TO THE 2005 PARALYTIC SHELLFISH POISONING EVENT IN NEW ENGLAND

From May to July 2005, the most severe bloom since 1972 of the microscopic toxic alga, *Alexandrium fundyense*, spread from Maine to Massachusetts, resulting in extensive commercial and recreational shellfish- harvesting closures to protect people from paralytic shellfish poisoning (PSP). Previous and current funding by a variety of programs, including the NSF and NOAA ECOHAB and MERHAB programs, the NCCOS Event Response, and the NSF-NIEHS WHCOHH (see Box 1), along with critical information provided by the Gulf of Maine Ocean Observing System buoys, allowed scientists in the region to provide maps of the bloom location and model predictions of bloom movement to management agencies responsible for ensuring safe shellfish. These information products helped state agencies focus limited resources on monitoring areas where new closures were needed or where reopening might be possible. The FDA and NOAA also used this information to make decisions regarding closure of offshore areas under federal jurisdiction. Thus, use of state-of-the-art field and laboratory techniques by OHH researchers allowed the bloom to be monitored in real time, enabling early warnings to public health and fishery managers. As a result of the coordinated assistance given by researchers at Woods Hole Oceanographic Institution, FDA, and NOAA, there were no reported human PSP illnesses, despite remarkably high toxicity in unmarketed shellfish products during this bloom.

4: IMPLEMENTING A STRONG INTERAGENCY OHH RESEARCH PROGRAM

In developing this plan, the IWG-4H had extensive deliberations about the comprehensive lists of research, infrastructure and outreach priorities that comprise Chapter 3. Based on those discussions, and numerous interactions with OHH researchers and others during the information gathering efforts that resulted in those priorities, the IWG-4H distilled the following six implementation actions to ensure a vibrant Interagency OHH Program. These actions, and the priorities identified within them, will advance OHH research across a broad interdisciplinary and interagency front and transfer and apply OHH findings to reduce human health risks, maximize human health benefits, and ensure healthy and productive marine ecosystems.

Action 1: Work through Existing OHH Programs and Partnerships

The OHH Act of 2004 stated that this Implementation Plan should "build on and complement the ongoing activities of the National Oceanic and Atmospheric Administration, the National Science Foundation, and other departments and agencies ..." The existing formal programs encompass NOAA Centers of Excellence in OHH, the NSF-NIEHS Centers for OHH, and the NOAA OHHI's extramural research grants, traineeships, and distinguished scholars. Other agencies included in this plan are CDC, EPA, FDA, MMC, NASA, and USGS. Efforts of all these collaborating agencies and their current and future partners to conduct research targeted at the high priority topics (see Action 2 below), support OHH-related infrastructure (see Action 3), and conduct outreach, education, and application activities (see Action 4) will increase the ability of the Interagency OHH Program to reduce ocean-based risks to humans and develop new marine-derived products to improve human well-being.

Action 2: Target Priority Research

Based on the numerous major opportunities for advancement of OHH science described in Chapter 3, and the respective interests, roles, and priorities of the various agencies as outlined in Table 3, highest priority should be given to six research areas:

Basic and applied studies of ocean ecosystems and processes that affect human exposure to ocean- related health risks, including development of new and improved sensors, assays, tools and methods for the assessment, monitoring and prediction of HABs and their toxins, known and potentially infectious diseases (including those that may be transmitted from marine animals to humans and from humans to animals), and ECCs. Such investigations should include assessments of the impacts of natural hazards, seafood consumption, and climate change on human exposure to health risks from the ocean and ways to reduce, mitigate or prevent such impacts.

Discovery and development of new marine pharmaceuticals and other products beneficial to humans. The ocean is the last frontier for new drug discovery, and because of its unparalleled biodiversity, it holds huge possibilities for new products that could improve and perhaps even revolutionize health care. Efforts should optimize acquisition, testing, characterization, and production of new marine bioproducts, including those produced through sustainable aquaculture. Work in this target area should emphasize ecologically sound methods and innovative public-private collaborations. Because of the lengthy lag time between initial discovery and product development, this work requires a long-term focus and commitment.

Epidemiological studies and disease surveillance, both short- and long-term, are necessary to document and elucidate acute and chronic health effects in humans and animals resulting from exposures to ocean waters, marine sediments, aerosols, and seafood. These investigations include collaboration with entities collecting environmental data and the integration of environmental and health data to document current and future levels of human disease resulting from ocean-based exposures, increase understanding of relationships among environmental factors and distribution/abundance of harmful agents, develop predictive models, and identify promising means to limit and prevent human exposures to these agents. This work also encompasses disease surveillance in marine mammals, seabirds, and other marine animals likely to harbor pathogens of potential danger to humans.

Use of marine species and habitats as early warning sentinels to indicate existing and emerging threats to human, animal and ecosystem health **and as models** for the study of human disease processes and toxicology.

Improving, developing, and testing conceptual and quantitative models that integrate a broad range of environmental, biological, and epidemiological data and produce new levels of understanding and predictive capacity related to ocean health threats.

Economic and socio-cultural studies that advance understanding of how humans use and value the health benefits provided by coastal and ocean waters and resources and how these uses and values are impacted by real and perceived health threats that these environments may pose.

Table 3. Primary Roles of Agencies in Conducting and Utilizing Research in Priority Areas Identified in the Interagency Oceans and Human Health Research implement-tation Plan (see Chapter 2)

Research Topic/Agency	NOAA	NSF	NIEHS	CDC	EPA	FDA	MMC	NASA	USGS
Pathogens	*RTSA	*RT	*RTS	RTS	*TSA	*RTSA	RA		RT
Chemical Contaminants	*RTSA	RT	RTS	A	*RTA	*RTSA	RA		RTS
Harmful Algal Blooms	*RTSA	*RT	*RTSA	*RTS	*RTA	*RTSA	RA	*RS	RT
Seafood Safety	*RTSA		RSA	S		*RTSA	A		
Pharmaceuticals & Other Beneficial Products	*RTA	*RT	*RT		*TA	*RTSA			
Epidemiology	*RTSA		RTSA	*RTSA	*RTS	*SA	A		
Sentinels	*RTSA			S		*RTSA	RA		
Genomics & Other "Omic" Sciences	*RTA	*RT	*RTS		*RT	*RTSA			
Social, Behavioral, Economic	*RTA	R	R		*R	*SA	A		
Climate Change	*RTSA	*RT		RT	*RSA		*RSA	*RS	RTS

R = Research, basic to applied (mission-specific)
T = Tool and/or method development
S = Surveillance and monitoring
A = Application (use) and/or regulation
* = Current activity within OHH-related program

Action 3: Support Research Infrastructure

Like almost all interdisciplinary research efforts, the Interagency OHH Program requires the right mix of cutting-edge research and supporting infrastructure. Based on an assessment of the respective interests, roles, and priorities of the agencies participating in the OHH program (Table 4), supporting infrastructure includes:

The IOOS and other ocean observatories that contribute to the Global Earth Observing System of Systems. These observing systems are essential building blocks for synoptic collection of physical, chemical, and biological data related to ocean- based health threats, through sensors and other tools, and for integrating these data into useful and timely information products for decision-makers.

Computing, data management, and bioinformatic infrastructure to enable data sharing, integration, archiving, analysis, and access for a broad range of OHH-related data, including the large amounts of data derived from genomic and proteomic studies.

New standards and standardized methods for OHH research and research materials. This activity should include capabilities for sample and specimen archiving, including those obtained for new product discovery, and the rapid evaluation, validation, and implementation of new analytical and monitoring methods by researchers and federal regulatory and management agencies.

Access platforms that make ocean sampling, observations, and discovery possible, including research vessels, satellites, aircraft, buoys, AUVs, ROVs, and others, and dedication of sufficient time on these platforms for the OHH program to accomplish crucial sampling activities.

Core facilities in genomics/proteomics, marine microbiology and analytical chemistry. Sequencing centers and state-of-the-art microbiology laboratories, with associated computer processing, bioinformatics, and analytical capabilities, support "omics" research to advance understanding of health impacts and disease processes at the molecular level and the identification and quantification of pathogenic microbes. In addition, chemistry laboratories with the capacity to provide consistent measurements of an increasing number and variety of chemical pollutants at very low but biologically relevant concentrations (e.g., levels < 1 part per trillion) in water, sediment, air, and tissue samples will facilitate our efforts to address the associated human health concerns. These core facilities will allow rapid progress that may transform our understanding of disease processes and potential control, treatment, and prevention options.

Action 4: Support Transition of OHH Research to Application

To fulfill the vision of the Interagency OHH Program, OHH research, information, and technology must be effectively transferred to the OHH research community, resource and public health managers, other users, and the public. Also, a national OHH research program will require training a new generation of scientists who are comfortable and equipped to work at the juncture of ocean and biomedical sciences and who are able to communicate with multiple and very different users. Based on an assessment of the respective interests, roles, and priorities of the agencies participating in the OHH program (Table 5), targeted and coordinated information, outreach, and education activities across the Interagency OHH Program should:

Strengthen cross-agency, cross-institution, and cross-discipline collaborations that provide the necessary framework to develop a fundamentally new interdisciplinary research community with the breadth of expertise required to address diverse human health questions within the complexity of ocean systems and ensure coordination of agency OHH-related communication and education efforts.

Provide opportunities for interdisciplinary training, research, and collaborations for graduate and post-doctoral students, health professionals, scientists at early- to mid-career stages, and scholars that will expose them to both ocean and health sciences and enable them to integrate data and approaches across fields and effectively communicate and partner with multiple user groups.

Table 4. Primary Roles/Interests of Agencies in Providing and/or Using Major Infrastructure for the Interagency Oceans and Human Health Research Implementation Plan (see Chapter 2)

Agency/Infrastructure	NOAA	NSF	NIEHS	CDC	EPA	FDA	MMC	NASA	USGS
Link to Observing Systems	*HP PU	*HP P	*HP U	HP PU	*HP	HP PU	HP U	*HP PU	HP P
Data Management, Access & Modeling	*HP PU	*HP P	*HP PU	HP PU	*HP	HP PU	HP U	*HP PU	HP P
Development of Standards	MP PU	*MP PU	MP PU	HP U	LP	HP PU	MP U	MP PU	MP PU
Access to Sea	*HP PU	*HP P	*HP PU	LP U	LP	HP U	LP U	MP U	MP U

HP = High Priority; MP = Medium Priority; LP = Low Priority .
* = ongoing activity within the agency's OHH-related efforts
P = Provider and U = User; any agency could be one or the other or both

Provide OHH information, tools, technology, products, and training to support improved public health and ocean resource decision-making. A robust OHH research program, coupled with monitoring and surveillance, data sharing, and technology transfer, will lead to availability of a host of helpful products for use by decision-makers and the public. A targeted outreach effort will allow the OHH program to strengthen partnerships with managers and decision-makers, and effectively communicate with them and respond to their information and technology needs on a rapid, regular, and reliable basis.

Coordinate OHH communication programs across agencies. Although outreach is a particular responsibility of the NOAA OHHI, OHH information programs should be coordinated among mission and science agencies at all levels of government to enable rapid transfer and consistent communication of findings and tools to users, including a nationwide network of health care providers. This effort should encompass potential or emerging threats detected from human and animal disease surveillance, monitoring for safety of seafood and recreational and drinking waters, and identification of ocean-related health benefits.

Support interagency partnerships to enable rapid and coordinated response of OHH researchers to emergency situations such as may be associated with severe weather events, sewage, oil or chemical spills, and possible marine-derived disease outbreaks among both humans and animals. Findings must then be communicated effectively to the public.

Promote ocean stewardship and ocean and human health literacy by improving the visibility and public awareness of the relationships between oceans and human health and introducing OHH themes, activities, and materials within schools and in a variety of settings such as aquariums, zoos, and via the popular press and media.

Action 5. Improve Coordination of OHH Activities within and across Agencies and Internationally

The OHH Act requires that "The President, through the National Science and Technology Council, shall coordinate and support a national program to improve understanding of the role of the oceans in human health." This program is to be carried out in consultation with the Interagency Task Force on Harmful Algal Blooms and Hypoxia established under section 603 of the Harmful Algal Blooms and Hypoxia Control Act of 1998 (HABHRCA). Currently, responsibilities for the duties of the Interagency Task Force on Harmful Algal Blooms and Hypoxia have been assigned to the IWG-4H, which is also responsible for development of this Interagency OHH Research Implementation Plan and the Interagency OHH Annual Report.

Table 5. Primary Roles/Interests of Agencies in Transitioning OHH Research Results to Applications, Outreach and Education (see Chapter 2)

Agency/Transitions	NOAA	NSF	NIEHS	CDC	EPA	FDA	MMC	NASA	USGS
Outreach									
Transition of research results to products or applications	HP		*MP	HP	LP	*HP	MP	*HP	*HP
OHH advisory information to public health officials, resource managers & general public	*HP		*MP	*HP	*HP	*HP	MP	LP	*MP
Education									
Improve public understanding of ocean sciences, including OHH	*MP	*HP	*MP	LP	LP	*HP	MP	*HP	*MP
Professional training of undergraduate, graduate students & post-doctoral researchers	*HP	*HP	*HP	LP	LP	*MP	MP	*HP	LP
Rapid Response									
Assess health risks	*HP		*HP	*HP	*HP	*HP	HP		*MP
Follow up research	*HP	*HP	*HP	*HP	LP	*HP	HP	HP	*MP
Provide analytical resources	*HP	*MP	*HP	*HP	LP	*HP	HP		*HP

HP = High Priority; MP = Medium Priority; LP = Low Priority
* = ongoing activity within the agency's OHH-related efforts

Much integration has been achieved on an ad hoc basis by the program leaders and participants in the NOAA and NSF-NIEHS OHH programs and through the work to date of the IWG-4H. The JSOST, through the IWG4H, will continue to advance interagency coordination.

Action 6. Provide for Updates to the Interagency OHH Research Implementation Plan

The OHH Act of 2004 requires that the Interagency OHH Annual Report include a summary or copy of the Implementation Plan and any changes made in the plan. Thus, this annual reporting mechanism provides a regular opportunity for the interagency OHH community to update the Implementation Plan. Programmatic support for interagency-sponsored workshops that would include the external research and stakeholder communities, as well as federal agencies, at 3-5 year intervals to examine progress and re-evaluate the Plan would also be beneficial.

APPENDIX 1: OCEANS AND HUMAN HEALTH ACT OF 2004

TITLE IX-OCEANS AND HUMAN HEALTH ACT

Sec. 901. Short Title

This title may be cited as the ‘Oceans and

Sec. 902. Interagency Oceans and Human Health Research Program

(a) COORDINATION-The President, through the National Science and Technology Council, shall coordinate and support a national research program to improve understanding of the role of the oceans in human health.

(b) IMPLEMENTATION PLAN-Within 1 year after the date of enactment of this Act, the National Science and Technology Council, through the Director of the Office of Science and Technology Policy shall develop and submit to the Congress a plan for coordinated Federal activities under the program. Nothing in this subsection is intended to duplicate or supersede the activities of the Inter-Agency Task Force on Harmful Algal Blooms and Hypoxia established under section 603 of the Harmful Algal Bloom and Hypoxia Research and Control Act of 1998 (16 U.S.C. 1451 note). In developing the plan, the Committee will consult with the Inter-Agency Task Force on Harmful Algal Blooms and Hypoxia. Such plan will build on and complement the ongoing activities of the National Oceanic and Atmospheric Administration, the National Science Foundation, and other departments and agencies and shall-

(1) establish, for the 10-year period beginning in the year it is submitted, the goals and priorities for Federal research which most effectively advance scientific understanding of the connections between the oceans and human health, provide usable information for the prediction of marine-related public health problems and use the biological potential of the oceans for development of new treatments of human diseases and a greater understanding of human biology;

(2) describe specific activities required to achieve such goals and priorities, including the funding of competitive research grants, ocean and coastal observations, training and support for scientists, and participation in international research efforts;
(3) identify and address, as appropriate, relevant programs and activities of the Federal agencies and departments that would contribute to the program;
(4) identify alternatives for preventive unnecessary duplication of effort among Federal agencies and departments with respect to the program;
(5) consider and use, as appropriate, reports and studies conducted by Federal agencies and departments, the National Research Council, the Ocean Research Advisory Panel, the Commission on Ocean Policy and other expert scientific bodies;
(6) make recommendations for the coordination of program activities with ocean and human health-related activities of other national and international organizations; and
(7) estimate Federal funding for research activities to be conducted under the program.

(c) PROGRAM SCOPE- The program may include the following activities related to the role of oceans in human health:
(1) Interdisciplinary research among the ocean and medical sciences, and coordinated research and activities to improve understanding of processes within the ocean that may affect human health and to explore the potential contribution of marine organisms to medicine and research, including—
(A) vector- and water-borne diseases of humans and marine organisms, including marine mammals and fish;
(B) harmful algal blooms and hypoxia (through the Inter-Agency Task Force on Harmful Algal Blooms and Hypoxia);
(C) marine-derived pharmaceuticals;
(D) marine organisms as models for biomedical research and as indicators of marine environmental health;
(E) marine environmental microbiology;
(F) bioaccumulative and endocrine-disrupting chemical contaminants; and
(G) predictive models based on indicators of marine environmental health or public health threats.
(2) Coordination with the National Ocean Research Leadership Council (10 U.S.C. 7902(a)) to ensure that any integrated ocean and coastal observing system provides information necessary to monitor and reduce marine public health problems including health-related data on biological populations and detection of contaminants in marine waters and seafood.
(3) Development through partnerships among Federal agencies, States, academic institutions, or non-profit research organizations of new technologies and approaches for detecting and reducing hazards to human health from ocean sources and to strengthen understanding of the value of marine biodiversity to biomedicine, including—
(A) genomics and proteomics to develop genetic and immunological detection approaches and predictive tools and to discover new biomedical resources;

(B) biomaterials and bioengineering;
(C) in situ and remote sensors used to detect, quantify, and predict the presence and spread of contaminants in marine waters and organisms and to identify new genetic resources for biomedical purposes;
(D) techniques for supplying marine resources, including chemical synthesis, culturing and aquaculturing marine organisms, new fermentation methods and recombinant techniques; and
(E) adaptation of equipment and technologies from human health fields.

(4) Support for scholars, trainees and education opportunities that encourage an interdisciplinary and international approach to exploring the diversity of life in the oceans.

(d) ANNUAL REPORT- Beginning with the first year occurring more than 24 months after the date of enactment of this Act, the National Science and Technology Council, through the Director of the Office of Science and Technology Policy shall prepare and submit to the President and the Congress not later than January 31st of each year an annual report on the activities conducted pursuant to this title during the preceding fiscal year, including—

(1) a summary of the achievements of Federal oceans and human health research, including Federally supported external research, during the preceding fiscal year;
(2) an analysis of the progress made toward achieving the goals and objectives of the plan developed under subsection (b), including identification of trends and emerging trends;
(3) a copy or summary of the plan and any changes made in the plan;
(4) a summary of agency budgets for oceans and human health activities for that preceding fiscal year; and
(5) any recommendations regarding additional action or legislation that may be required to assist in achieving the purposes of this title.

Sec. 903. National Oceanic and Atmospheric Administration Oceans and Human Health Initiative

(a) ESTABLISHMENT- As part of the Interagency Oceans and Human Health Research Program, the Secretary of Commerce is authorized to establish an Oceans and Human Health Initiative to coordinate and implement research and activities of the National Oceanic and Atmospheric Administration related to the role of the oceans, the coasts, and the Great Lakes in human health. In carrying out this section, the Secretary shall consult with other Federal agencies conducting integrated oceans and human health research and research in related areas, including the National Science Foundation. The Oceans and Human Health Initiative is authorized to provide support for—

(1) centralized program and research coordination;
(2) an advisory panel;
(3) one or more National Oceanic and Atmospheric Administration national centers of excellence;

(4) research grants; and
(5) distinguished scholars and traineeships.

(b) ADVISORY PANEL- The Secretary is authorized to establish an oceans and human health advisory panel to assist in the development and implementation of the Oceans and Human Health Initiative. Membership of the advisory group shall provide for balanced representation of individuals with multi-disciplinary expertise in the marine and biomedical sciences. The Federal Advisory Committee Act (5 U.S.C. App.) shall not apply to the oceans and human health advisory panel.

(c) NATIONAL CENTERS-

(1) The Secretary is authorized to identify and provide financial support through a competitive process to develop, within the National Oceanic and Atmospheric Administration, for one or more centers of excellence that strengthen the capabilities of the National Oceanic and Atmospheric Administration to carry out its programs and activities related to the oceans' role in human health.

(2) The centers shall focus on areas related to agency missions, including use of marine organisms as indicators for marine environmental health, ocean pollutants, marine toxins and pathogens, harmful algal blooms, hypoxia, seafood testing, identification of potential marine products, and biology and pathobiology of marine mammals, and on disciplines including marine genomics, marine environmental microbiology, ecological chemistry and conservation medicine.

(3) In selecting centers for funding, the Secretary will give priority to proposals with strong interdisciplinary scientific merit that encourage educational opportunities and provide for effective partnerships among the Administration, other Federal entities, State, academic, non-profit research organizations, medical, and industry participants.

(d) EXTRAMURAL RESEARCH GRANTS-

(1) The Secretary is authorized to provide grants of financial assistance to the scientific community for critical research and projects that explore the relationship between the oceans and human health and that complement or strengthen programs and activities of the National Oceanic and Atmospheric Administration related to the ocean's role in human health. Officers and employees of Federal agencies may collaborate with, and participate in, such research and projects to the extent requested by the grant recipient. The Secretary shall consult with the oceans and human health advisory panel established under subsection (b) and may work cooperatively with other agencies participating in the interagency program to establish joint criteria for such research and projects.

(2) Grants under this subsection shall be awarded through a competitive peer-reviewed, merit-based process that may be conducted jointly with other agencies participating in the interagency program.

(e) TRAINEESHIPS- The Secretary of Commerce is authorized to establish a program to provide traineeships, training, and experience to pre-doctoral and post-doctoral students and to scientists at the beginning of their careers who are interested in the oceans in human health research conducted under the NOAA initiative.

Sec. 904. Public Information and Outreach

(a) IN GENERAL- The Secretary of Commerce, in consultation with other Federal agencies, and in cooperation with the National Sea Grant program, shall design and implement a program to disseminate information developed under the NOAA Oceans and Human Health Initiative, including research, assessments, and findings regarding the relationship between oceans and human health, on both a regional and national scale. The information, particularly with respect to potential health risks, shall be made available in a timely manner to appropriate Federal or State agencies, involved industries, and other interested persons through a variety of means, including through the Internet.

(b) REPORT- As part of this program, the Secretary shall submit to Congress an annual report reviewing the results of the research, assessments, and findings developed under the NOAA Oceans and Human Health Initiative, as well as recommendations for improving or expanding the program.

Sec. 905. Authorization of Appropriations

There are authorized to be appropriated to the Secretary of Commerce to carry out the National Oceanic and Atmospheric Administration Oceans and Human Health Initiative, $60,000,000 for fiscal years 2005 through 2008. Not less than 50 percent of the amounts appropriated to carry out the initiative shall be utilized in each fiscal year to support the extramural grant and traineeship programs of the Initiative.

APPENDIX 2: AGENCY PROSPECTUSES

The National Science Foundation (NSF) is an independent federal agency with a primary mission to promote the progress of science; to advance the national health, prosperity, and welfare; and to secure the national defense. NSF provides approximately 20 percent of all federally supported basic research conducted by America's colleges and universities. In many fields such as mathematics, computer science and the social sciences, NSF is the major source of federal backing.

NSF supports research efforts focused on Oceans and Human Health in an interagency partnership with the National Institute of Environmental Health Sciences (NIEHS). Four Centers for Oceans and Health (COOH) were established in 2004. The NSF-NIEHS COOH effort is managed in the Division of Ocean Sciences (OCE) by the Chemical Oceanography Program. NSF also participates in other Interagency activities with OHH themes, including: Ecology and Oceanography of Harmful Algal Blooms (ECOHAB) (with NOAA, EPA, ONR, NASA); Ecology of Infectious Diseases (EID) (with NIH). NSF also supports basic research that addresses OHH Themes in various Directorates of the Foundation. Specific themes include harmful algal blooms (HABs) (Directorate of Geosciences/Division of Ocean Sciences), microbial ecology (primarily in the Directorates of Biology and Geosciences). The Directorate of Social, Behavioral, and Economic Sciences (SBE) also has identified the human dimensions of OHH as a future area of interest. The Division of Ocean Sciences supports education and outreach efforts in the ocean sciences through Centers for Ocean Science Education Excellence (COSEE) Program. Sea-going capability in support of academic research is primarily provided by the University-National Oceanographic Laboratory System (UNOLS). NSF provides approximately 70 percent of support for UNOLS operations. Oversight and support for UNOLS is provided by the Division of Ocean Sciences. The Foundation will implement two "Major Research Equipment and Facilities Construction (MREFC)" projects, the Ocean Observatories Initiative (OOI); and the National Ecological Observatory Network (NEON). These will provide research observatory capabilities in the marine and freshwater/terrestrial environments, respectively, and thus have a high potential to contribute to the developing OHH activities. The OOI is expected to be come operational in the coastal regions in the 2008-2012 time frame.

The Centers for Disease Control and Prevention (CDC) is one of the 13 major operating components of the Department of Health and Human Services (HHS). HHS is the principal agency in the United States government for protecting the health and safety of all Americans and for providing essential human services, especially for those people who are least able to help themselves. Since it was founded in 1946 to help control malaria, CDC has remained at the forefront of public health efforts to prevent and control infectious and chronic diseases, injuries, workplace hazards, disabilities, and environmental health threats. Today, CDC is globally recognized for conducting research and investigations and for its action-oriented approach. CDC applies research and findings to improve people's daily lives and responds to health emergencies—something that distinguishes CDC from its peer agencies. CDC supports research efforts related to OHH at many levels and through several internal units, including the Health Studies Branch at the National Center for Environmental Health (NCEH), the National Center for Infectious Diseases (NCID), and the National Institute for Occupational Safety and Health (NIOSH).

APPENDIX 3: LIST OF ACRONYMS AND ABBREVIATIONS

Act	Oceans and Human Health Act
ALS/PDC	Amyotrophic Lateral Sclerosis/Parkinsonism–Dementia Complex
AMAP	Artic Marine Assessment Program
AOAC	Association of Chemists (now AOAC International)
ARISA	Advanced Reflectometer Surface Analysis
AUV	Autonomous Underwater Vehicle
AZT	Azidothymidine (an antiretroviral drug)
BMAA	ß-N-methylamino-L-alanine (a neurotoxin amino acid)
CA	California
CAMERA	Community Cyberinfrastructure for Advanced Marine Microbial Ecology Research and Analysis
CDC	Centers for Disease Control and Prevention
CDER	Center for Drug Evaluation and Research within FDA

CEGLHH	NOAA Center of Excellence for Great Lakes and Human Health
CFSAN	Center for Food Safety and Applied Nutrition within FDA
COHH	Centers for Oceans and Human Health within NSF
COSEE	Centers for Ocean Science Education Excellence
CSDM	Content Standard for Digital Geospatial Metadata
CTX-H	a ciguatoxin congener derived from the dinoflagellate *Gambierdiscus*
DMAC	Data Management and Communications Plan
DNA	Deoxyribonucleic Acid
ECC	Emerging Chemicals of Concern
ECOHAB	Ecology and *Oceanography* of Harmful Algal Blooms (multi-agency program involving NSF, NOAA, EPA, ONR, NASA)
EEZ	Exclusive Economic Zone
ELISA	Enzyme-Linked Immunosorbent Assay
EPA	Environmental Protection Agency
FAO	Food and Agriculture Organization of the United Nations
FDA	Food and Drug Administration
FGDC	Federal Geographic Data Committee
FLDH	Florida Department of Health
GA	Georgia
GEO	Group on Earth Observations
GEOSS	Global Earth Observation System of Systems
GIS	Geographic Information System
GLERL	Great Lakes Environmental Research Laboratory (a NOAA Center of Excellence for OHH)
GRACE	Gravity Recovery and climate Experiment within NASA
GSAT	Global Situational Tool
HAB	Harmful Algal Bloom
HABISS	Harmful Algal Bloom-related Illness Surveillance System
HABHRCA	Harmful Algal Bloom and Hypoxia Control Act of 1998
HABSOS	NOAA's Harmful Algal Blooms Observing System
HACCP	Hazard Analysis Critical Control Point
HARR-HD	A report entitled, "Harmful Algal Research and Response: A Human Dimensions Strategy"
HARRNESS	Harmful Algal Research and Response National Environmental Science Strategy, 2005-2015
HHS	Department of Health and Human Services
HIV	Human Immunodeficiency Virus
HML	Hollings Marine Laboratory (NOAA Center of Excellence for Oceans and Human Health at the Hollings Marine Laboratory)
HTC	High Throughput Culturing
IAEA	International Atomic Energy Agency
ICES	International Council for the Exploration of the Seas

ICOSRMI	Interagency Committee on Ocean Science and Resource Management Integration
IGBP	International Geosphere-Bioshere Programme
IOC	International Oceanographic Commission
IOM	Institute of Medicine of the National Academies
IOOS	Integrated Ocean Observing System
ISSC	Interstate Shellfish Sanitation Conference
IWC	International Whaling Commission
IWG-4H	Interagency Working Group on Harmful Algal Blooms, Hypoxia, and Human Health
JSOST	Joint Subcommittee on Ocean Science and Technology
LA	Los Angeles
LiDAR	Light Detection and Ranging; or Laser Imaging Detection and Ranging
Mass.	Massachusetts
MERHAB	Monitoring and Event Response for Harmful Algal Blooms program
MMC	Marine Mammal Commission
MML	Mote Marine Laboratory
MODIS	Moderate Resolution Imaging Spectroradiometer
MYP	Multi-Year Plan
NASA	National Aeronautics and Space Administration
NASA ROSES	NASA's Research Opportunities in Space and Earth Sciences
NC	North Carolina
NCCR	National Coastal Condition Report
NCEH	National Center for Environmental Health within CDC
NCID	National Center for Infectious Diseases within CDC
NEON	National Ecological Observatory Network within NSF
NIEHS	National Institute of Environmental Health Science within NIH
NIH	National Institutes of Health
NIST	National Institute of Standards and Technology
NOAA	National Oceanic and Atmospheric Administration
NOAA HABSOS	NOAA's Harmful Algal Blooms Observing System
NOAA OHHI Centers	NOAA'S Oceans and Human Health Initiative Centers of Excellence
NOPP	National Oceanographic Partnership Program
NOS	National Oceans Service within NOAA
NRC	National Research Council
NRL	Naval Research Laboratory
NSF	National Science Foundation
NSTC	National Science and Technology Council
NWFSC	Northwest Fisheries Science Center
NURP	National Underwater Research Program
OARSA	Office of Applied Research and Safety Assessment within FDA

OE	Ocean Exploration Program
OHH	Oceans and Human Health
OHH	Act Oceans and Human Health Act of 2004
OHHI	Oceans and Human Health Initiative
OIE	World Organization for Animal Health
ONR	Office of Naval Research
OOI	Ocean Observatories Initiative (within NSF)
OR	Oregon
ORD	Office of Research and Development withn in EPA
OS	Office of Seafood within FDA
OSV	Ocean Survey Vessel
PAH	Polycyclic Aromatic Hydrocarbon
PBDE	Polybrominated Diethyl Ether
PCB	Polychlorinated Biphenyls
PICES	North Pacific Marine Science Organization
PNW	Pacific Northwest
PNW H2O	Pacific Northwest Center for Oceans and Human Health (funded by NSF-NIEHS at the University of Washington)
PRCMB	University of Hawaii Pacific Research Center for Marine Biomedicine (funded by NSF-NIEHS)
PSP	P aralytic Shellfish Poisoning
QA	Quality Assurance
RA	Regional Association
REASoN	Research, Education and Applications
Solution	Network
REU	Research Experiences for Undergraduate Students
RFA	Request for Applications
RFP	Request for Proposals
Rho-GTP	A protein involved with functioning of the nucleotide guanosine triphosphate important in metabolism
ROV	Remotely Operated Vehicle
R/V	Research Vessel
SAFE	Swim Advisory Forecast Estimate
SARS	Severe Acute Respiratory Syndrome
SBE	Social, Behavioral, and Economic
SC	South Carolina
SEAPORT	Signal Environmental and Plankton Observations in Real Time
SEPMN	Southeast Phytoplankton Monitoring Network within NOAA
SIVCA	Species Identification via Chimeric Amplification
sp.	species (singular)
spp.	species (plural)
START	Solutions to Avoid Red Tide
TOS	The *Oceanography* Society
TRMM	Tropical Rainfall Measuring Mission with-in NASA
UMCOHH	University of Miami Center for Subtropical and Tropical

	Oceans and Human Health Research in the Marine Sciences (funded by NSF-NIEHS)
UNESCO	United Nations Educational, Scientific, and Cultural Organization
UNOLS	University-National Oceanographic Laboratory System
U.S.	United States
USCOP	U.S. Commission on Ocean Policy
USDA	U.S. Department of Agriculture
USFWS	U.S. Fish and Wildlife Service
USGS	U.S. Geological Survey
UW	University of Washington
WA	Washington
WCCOHH	NOAA West Coast Center of Excellence for Oceans and Human Health at the Northwest Fisheries Science Center
WHCOHH	Woods Hole Center for Oceans and Human Health (funded by NSF-NIEHS)
WHO	World Health Organization
WHOI	Woods Hole Oceanographic Institute

Appendix 4: Joint Subcommittee on Ocean Science and Technology (JSOST)

Julie Morris, NSF, Acting Co-Chair
Richard Spinrad, DOC/NOAA, Co-Chair
Dan Walker, OSTP, Co-Chairs

Arctic Research Commission
John Farrell

Department of Agriculture
Louie Tupas
Alternate: Meryl Broussard

Department of Commerce
National Oceanic and Atmospheric Administration
Steve Murawski
Alternate: Marie Colton

Department of Defense
Charles Chesnutt
Frank Herr
Alternate: Joan Pope
Alternate: James E. Eckman

Department of Energy
Jerry Elwood
Alternate: James Ahlgrimm

Department of Health and Human Services
Allen Dearry
William Jones
Michael McGeehin
Alternate: Lorraine Backer
Alternate: G. David Williamson

Department of Homeland Security
Jonathan Berkson

Department of the Interior
James Kendall
P. Patrick Leahy

Department of Justice
Bradford McLane

Department of State
David Balton
Alternate: Liz Tirpak

Department of Transportation
Todd Ripley

Environmental Protection Agency
George Gray
Alternate: Steven Hedtke

Executive Office of the President Council on Environmental Quality
Gerhard Kuska

Executive Office of the President Domestic Policy Council
Paul Skoczylas

Executive Office of the President Office of Management and Budget
Emily Woglom
Alternate: Kimberly Miller

Executive Office of the President Office of Science and Technology Policy
Dan Walker

Joint Chiefs of Staff
Robert Winokur
Alternate: Commander James Kraska

National Aeronautics and Space Administration
Jack Kaye
Alternate: Eric Lindstrom

National Science Foundation
Julie Morris

Marine Mammal Commission
Tim Ragen
David Laist

Smithsonian Institution
Leonard Hirsch
JSOST Ex Officio Members

JSOST Interagency Working Group on Facilities (IWG-F)
Robert Winokur, Chair, JCS

JSOST Interagency Working Group on Harmful Algal Blooms, Hypoxia and Human Health (IWG-4H)
Paul Sandifer, Co-Chair, NOAA
Lorrie Backer, Co-Chair, CDC

JSOST Interagency Working Group on Ocean and Coastal Mapping (IWG-OCM)
John Haines, Chair, USGS

JJSOST Interagency Working Group on Ocean Observations (IWGOO)
Richard Spinrad, Chair, NOAA

JSOST Interagency Working Group on Ocean Partnerships (IWG-OP)
Jim Kendall, Co-Chair, MMS
Eric Lindstrom, Co-Chair, NASA

Ocean Research and Resources Advisory Panel (ORRAP)
Ellen Prager, Chair
Stephen Weisberg, Vice-Chair
Debra Hernandez, Vice-Chair

Subcommittee for Integrated Management of Ocean Resources (SIMOR)
Mary Glackin, Co-Chair, NOAA
Chris Kearney, Co-Chair, DOI
Gerhard Kuska, Co-Chair, CEQ

Craig Hooks, Co-Chair, Acting, EPA

Subcommittee on Oceans Policy of the Global Environment Policy Coordinating Committee (Oceans Sub-PCC)
David Balton, Chair

REFERENCES AND WEB RESOURCES CONSULTED

Abraham, W. M. & Baden D. G. (2006). Case study: AerosolizedFlorida red tide toxins and human health effects. *Oceanography, 19*(2),107-*109.*

Aguirre, A. A., & Tabor G. M. (2004). Introduction: Marine ver- tebrates as sentinels of marine ecosystem health. *EcoHealth, 1*(3),236-*238.*

The Advisory Committee on Water Information and the NationalWater Quality Monitoring Council. (2006). *A national water quality monitoring network for U.S. coastal waters and their tributaries.* 99 p. http://acwi.gov/monitoring/network/design/

Alliance for Coastal Technologies (ACT). (2005). Genetic sensor forenvironmental water quality. St. Petersburg, FL. January 5-7, 2005. 19 p. http://www.act-us.info/download/workshop_reports/ ACT_WR05-01_Genetic_Sensors.pdf

Anderson, D. M., Hoagland, P. Kaoru Y. & White A. W. (2000). Estimated annual economic impacts from harmful algal blooms (HABs) in the United States. Technical Report WHOI-2000-11. Woods Hole Oceanographic Institution,Woods Hole, Mass.. 101 p. Anderson, D.M., B.A. Keafer, D.M. Kulis, R.M. Waters, and R. Nuzzi. 1993. An chrysophyte Aureococcus anophagefferens along the northeast coast of the United States. *Journal of Plankton Research 15*,563-*580.*

AOAC International: http://www.aoac.org/

ARMADA Project: http://www.armadaproject.org/

Atlantic States Marine Fisheries Commission (2007). The importance of habitat created by shellfish Coast of the U.S. Prepared by L.D. Coen and R. Grizzle with J. Lowery and K.T. Paynter, Jr., contributors. *South Carolina Department of Natural Resources, Marine Resources Division Educational Report 21.*

Backer, L. C. & McGillicuddy, Jr D. J. (2006). Harmful algalblooms: At the interface between coastal *Oceanography* and human health. *Oceanography, 19*(2),94-*106.*

Backer L. C., Schurz, R. H. Fleming, L. E. Kirkpatrick B. & Benson J. (2005). Phycotoxins in marine seafood. In W. Dabrowski (ed.), Chemical and functional properties of food components: toxins in food, p. 155-*190.* CRC Press, Boca Raton, FL.

Backer L., Fleming, L. E. Rowan A. & Baden D. (2003). Epidemiology and public health of human illnesses associated with harmful marine phytoplankton. In G.M. Hallegraeff, D.M. Anderson, and A.D. Cembella (eds.), Manual on harmful marine microalgae. UNESCO Monographs on Oceanographic Methodology, p. 723-750. UNESCO/WHO, Geneva, Switzerland.

Baden, D. G. (2005). Oceans and human health: A new era of environ- mental opportunities. In T. J. Goehl (ed.), Essays on the future of environmental health research: A tribute to Dr. Kenneth Olden, p 116-127. *Special Issue of Environmental Health Perspectives, Research Triangle Park, NC.*

Baden, D. G., Bourdelais, A. J. Jacocks, H. Michelliza & Naar J. (2005). Natural and derivative brevetoxins: Historical background, multiplicity, and effects. *Environmental Health Perspectives, 113*(5),621-625.

Baden, D. G., Fenical W. & Fuhrman J. (2001). Oceans and humanhealth roundtable report, 17-18 December, 2001. National Institute of Environmental Health Sciences and National Science Foundation, Research Triangle Park, NC. *Unpublished.* 27 p. http://www.niehs.nih.

Backer, L. C., Kirkpatrick, B. Fleming, L. E. Cheng, Y. S. Pierce, R.J. Bean, A. Clark, R. Johnson, D. Wanner, A. Tamer, R. Zhou, Y. & Baden D.G. (2005). Occupational exposure to aerosol-ized brevetoxins during Florida red tide events: Effects on a healthy worker population. *Environmental Health Perspectives, 113*(5),644-649.

Bauer, M. (ed.). 2006. Harmful algal research and response: A human dimensions strategy. Biotoxins and Harmful Algal Blooms, Woods Hole Oceanographic Institution, Woods Hole, Mass.. 58 p.

Bayley, S., Stotts, V. D. Springer P. F. & Steenis J. (1978). Changesin submerged aquatic macrophyte populations at the head of the survey of the brown tide Chesapeake Bay 1958-1975. *Estuaries, 1*,74-85.

Beach, D. (2002). Coastal sprawl: The effects of urban designon aquatic ecosystems in the United States. Pew Oceans Commission, Arlington, VA. 40 p. http://www.pewtrusts.org/pdf/env_pew_oceans_sprawl.pdf

Belden Russonello and Stewart and American Viewpoint (1999). and shell beds along theCommunicating about oceans: Results of a national survey. Conducted for The Ocean Project. 72 p. http://www.theocean-project

Bermuda time-series programs: http://bats.bbsr.edu/

Booth, S. & Zeller D. (2005). Mercury, food webs, and marinemammals: Implications of diet and climate change for human health. *Environmental Health Perspectives, 113*(5),521-526.

Bossart, G. D (2006). Case study: Marine mammals as sentinel spe- cies for oceans and human health. *Oceanography 19*(2):134-137.

Bowen, R. E., Frankic, A. & Davis. M. E. (2006). Human development and resource use in the health. *Oceanography, 19*(2),62-71.

Breslow, J. L (2006). N-3 fatty acids and cardiovascular disease. *American Journal of Clinical Nutrition, 83*(6),S1477-1482S.

Brian, J. V., Harris, C. A. Scholze, M. Backhaus, T. Booy, P. Lamoree, M. Pojana, G. Jonkers, N. Runnalls, T. Bonfa, A. Marcomini A. & Sumpter J. P. (2005). *Accurate prediction of the response of freshwater fish Environmental Health Perspectives, 113*(6),721-728.

Brodie, E. C., Gulland, F. M. D. Greig, D.J. Hunter, M. Jaakola, J. Leger, J. St. & Leighfield, T.A. Domoic acid causes reproductive failure in California sea lions (Zalophus californianus). *Marine Mammal Science, 22*(3),700-707.

Calder, P. C. (2006). N-3 -tion, and inflammatory Nutrition *83*(6):S1505-1519S.

Cane, M. A., Zebiak, S. E. & Dolan. S.C. (1986). Experimental fore- casts of El Nino. *Nature,* 321,827-832.

Chapman, R. W. (2001). EcoGenomics - a consilience for comparativeimmunology? *Developmental and Comparative Immunology, 25*(7),549-551.

Coen, L. D., Wilber, D. H. & Knott. D. In review. Development ofintertidal oyster reef resident communities on natural and constructed reefs in the southeastern U.S.: Can we restore native oyster habitats?. *Marine Ecology Progress Series.*

Coen, L. D., Luckenbach, M. W. & Breitburg. D. L. (1999). The role of oyster reefs as knowledge and some new perspectives. In L. R. Benaka (ed.), Fish habitat: Essential fish American Fisheries Society, Symposium 22, Bethesda, MD.

Cohen, J. T., Bellinger, D. C. Connor, W. E. Kris-Etherton, P. M. Lawrence, R. S. Savitz, D. A. Shaywitz, B. A. Teutsch, S. M. & Gray G. M. (2005). *A quantitative population fish consumption Medicine*, *29*(4),325.

Colgan, C. & Adkins. J. (2006). 2005 Hurricane damage to the Gulfof Mexico ocean economy. Monthly Labor Review. August, 2006. p. 76-78. http://www.bls.gov/opub/mlr/2006/08/art7full. pdf

Cox, P. A., Banack, S. A. Murch, S. J. Rasmussen, U. Tien, G. Bidigare, R. R. Metcalf, J. S. Morrison, L. F. Codd, G. A. & Bergman. B. (2005). Diverse taxa of cyanobacteria produce BMAA, a neurotoxic amino acid. *Proceedings of the National Academy of Sciences*, *102*,5074-5078.

Crossett, K. M., Culliton, T. J. Wiley, P. C. & Goodspeed. T. R. (2004). Population trends along the coastal Untied States: 1980-2008. National Oceanic and Atmospheric Administration. National Ocean Service. Special Projects. Silver Spring, MD. 47 p.

Dame, R. F. (1996). Ecology of marine bivalves: An ecosystemapproach. 254 p. CRC Marine Science Series, Boca Raton, FL.

Dame R, Bushek, D. & Prins. T. (2001). The role of suspension feeders as ecosystem transformers in shallow coastal environments. In: K. Reise (ed.), The ecology of sedimentary coasts, p. 11-37. Springer-Verlag, Berlin, Germany.

Daszak, P., Cunningham, A. A. & Hyatt. A. D. (2000). Emerging infectious diseases of wildlife- - Threats to biodiversity and human health. *Science*, *287*,443-449.

Data Management and Communications: http://dmac.ocean.us/dacsc/about.jsp;jsessionid=20308 16901163803430712

Data Management and Communications guidance: http://dmac.ocean.us/dacsc/guidance02.jsp;jsessionid=20301877 071164058782918

Daughton, C. G. & Ternes. T. A. (1999). Pharmaceuticals and personal care products in the environment: Agents of subtle change. Environmental Health Perspectives *107*(Supplement 6):907-938.

Dewailly, E. (2006). Case study: Canadian Inuit and the Arctic dilemma. *Oceanography, 19*(2),88-89.

Dewailly, E. & Knap. A. (2006). Food from the oceans and human health: Balancing risks and benefits. *Oceanography, 19*(2),84-93.

Dewailly, É. Furgal, C. Knap, A. Galvin, A. J. Baden, D. Bowen, B. Depledge, M. Duguay, L. Fleming, L. Ford, T. Moser, F. Owen, R. Suk,V & Unluata. U. (2002). Indicators of ocean and human health. Revue Canadienne de Sante Publique [*Canadian Journal of Public Health*], *93*(Suppl. 1),S34–38.

deWit, C. A. (2000). Brominated flame retardants. Swedish Environmental Protection Agency. Stockholm, Sweden. 94 p.

Dorfman, M., Stoner, N. & Merkel. M. (2004). Swimming in sewage. A report by the Natural Resources Defense Council and Environmental Integrity Project. 75 p. http://www.nrdc.org/ water/pollution/sewage/sewage.pdf

Dufour, A. P. & Wymer. L. J. (2006). Microbes, monitoring and human health. *Oceanography, 19*(2),72-80.

Dustan, P. & Halas. J. (1987). Changes in the reef-coral population of Carysfort Reef, Key Largo, Florida, 1974-1982. *Coral Reefs, 6*(2),91-106.

EPA water quality-based control program: http://www.epa.gov/waterscience/standards/

Eriksson, P., Jakobsson, E. & Fredriksson. A. (2001). Brominated flame retardants: A novel class of flame retardants in our environment. *Environmental Health Perspectives, 109*(9),903-908.

Federal Oceanographic Facilities Committee. (2001). Charting the Future for the National Academic Research Fleet. 32 p.

Fenical, W. (2006). Marine pharmaceuticals: Past, present, and future. *Oceanography, 19*(2),110-119.

Fleming L. E., Broad, K. Clement, A. Dewailly, E. Elmir, S. Knap, A. Pomponi, S.A. Smith, S. Solo Gabriele, H. & Walsh. P. (2006). Oceans and human health: Emerging public health risks in the marine environment. *Marine Pollution Bulletin, 53*,545-560.

Fleming, L. E. & Laws. E. (2006). Overview of the oceans and human health special issue. *Oceanography* 19(2):18-23.

Fleming L. E., Backer, L.C. & Baden. D. G. (2005). Overview of aerosolized Florida red tide toxins: Exposures and effects. *Environmental Health Perspectives, 113*(5),618-620.

Fleming, L. E., Kirkpatrick, B. Backer, L.C. Bean, J.A. Wanner, A. Dalpra, D. Tamer, R. Zaias, J. Cheng, Y.S. Pierce, R. Naar, J. Abraham, W. Clark, R. Zhou, Y. Henry, M.S. Johnson, D. Van De Bogart, G. Bossart, G.D. Harrington, M. & Baden. D. G. (2005). Initial evaluation of the effects of aerosolized Florida red tide toxins (brevetoxins) in persons with asthma. *Environmental Health Perspectives, 113*(5),650-657.

Fleming L. E., Backer, L. & Rowan. A. (2002). The epidemiology of human illnesses associated with harmful algal blooms. In D. Baden and D. Adams (eds.), Neurotoxicology handbook, volume *1*, p. 363-381. *Humana Press*, Inc., Totowa, *NJ*.

Flewelling, L. J., J. P. Naar, J. P. Abbbott, D. G. Baden, N. B. Barros, G. D. Bossart, M.-Y. D. Bottein, D. G. Hammond, E. M. Haubold, C. A. Heil, M. S. Henry, H. M. Jacocks, T. A. Leighfield, R. H. Pierce, T. D. Pitchford, S. A. Rommel, P. S. Scott, K. A. Steidinger, E. W. Truby, F. M. Van Dolah, & J. H. Landsberg. (2005). Brevetoxicosis: Red tides and marine mammal mortalities. *Nature, 435*,755-756.

Foran, J. A., Carpenter, D. O. Hamilton, M. C. Knuth, B. A. & Schwager. S. J. (2005). Risk-based consumption advice for farmed Atlantic and wild Pacific salmon contaminated with dioxins and dioxin-like compounds. *Environmental Health Perspectives, 113*(5),552-556.

Foster, I., Kesselman, C. & Tuecke. S. (2001). The Anatomy of the Grid: Enabling scalable virtual organization. *International Journal of High Performance Computing Applications, 15*(3),200-222.

Fulton, M. (1989). The effects of certain intrinsic and extrinsic variables on the lethal and sublethal toxicity of selected organophosphorus insecticides in the mummichog, Fundulus heteroclitus, under laboratory and field conditions. PhD Dissertation. *University of South Carolina.* 183 p.

Fulton, M. H., Scott, G. I Fortner, A. Bidleman, T. F. & Ngabe. B. (1993). The effects of urbanization on small high salinity estuaries of the southeastern United States. *Archives of Environmental Contamination and Toxicology, 25*,476-484.

Garces, E. & Anderson. D. M. (2005). SEED: A new EU-US collaboration on HABs. The Intergovernmental Oceanographic Commission of UNESCO. Harmful Algal News No. 28. p 1 and p 3. http://ioc.unesco.org/hab/news.htm.

GEOSS: http://www.earthobservations.org/index.html

Glibert, P. M., Anderson, D. Gentien, P. Graveli, E. & Sellner. K. (2005). The global, complex phenomena of harmful algal blooms. *Oceanography, 18*(2),36-147.

Given, S., Pendleton, L. H. & Boehm. A. B. (2006). Regional public health cost estimates of contaminated coastal waters: A case study of gastroenteritis at southern California beaches. *Environmental Science and Technology, 40*,4851 – 4858.

Global Ecosystem Dynamics databases: http://globec.whoi.edu/

Golbeck, J., Alford, A. & Hendler. J. (2004). Organization and structure of information using Semantic Web technologies. In R.W. Proctor and K-P.L. Vu(eds.), Handbook of human factors in web design, 752 p. Lawrence Ehlbaum Associates, Mahwah, NJ.

Grosell, M. & Walsh. P. J. (2006). Benefits from the sea: Sentinel species and animal models of human health. *Oceanography, 19*(2),126-133.

Goreau, T., McClanahan, T. Hayes, R. & Strong. A. (2000). Conservation of coral reefs after the 1998 global bleaching event. *Conservation Biology, 14*(1),5-15.

HACCP: http://www.cfsan.fda.gov/~lrd/haccp.html

Halevy, A. Y., Ives, Z. G. Tatarinov, I. & Mork. P. (2003). Piazza: Data management infrastructure for Semantic-Web applications. Proceedings of the Twelfth International World-Wide Web Conference, Budapest, Hungary, SESSION: Scaling up the Semantic Web. p. 556-567. http://www2003.org/cdrom/papers/ refereed/p540/p540-halevy.html

Hankin, S. & the DMAC Steering Committee. (2005). Data management and communications plan for research and operational Integrated Ocean Observing Systems: I. Interoperable data discovery, access, and archive. Ocean. U.S., Arlington, VA. 304 p. http://dmac.ocean.us/dacsc/docs/march2005_dmac_plan/dmac_ plan.pdf

HARRNESS. (2005). Harmful algal research and response: A national environmental science strategy 2005-2015. Randall, J.S., DM. Anderson, and P.M. Gilbert (eds.), Ecological Society of America. Washington, DC. 96 p.

Harvell, C. D., Mitchell, C. E. Ward, J.R. Altizer, S. Dobson, A.P. Ostfeld, R.S. & Samuel. M. D. (2002). Climate warming and disease risks for terrestrial and marine biota. *Science, 296*, 2158-2162.

Hawaii time series programs: http://hahana.soest.hawaii.edu/hot/ hot-dogs/

Hoagland, P. & Scatasta. S. (2006). The economic effects of harmful algal blooms. In E. Graneli and J. Turner (eds.), Ecology of harmful algae. Chapter 29. Ecology Studies Series. *SpringerVerlag*, Dordrecht, The Netherlands.

Hoagland, P., Anderson, D. M Kaoru, Y. & White. A.W. (2002). The economic effects of harmful algal blooms in the United States: Estimates, assessment issues, and information needs. *Estuaries, 25*(4b), 819–837.

Hodge, W. G., Schachter, H. M. Barnes, D. Pan, Y. Lowcock, E.C. Zhang, L. Sampson, M. Morrison, A. Tran, K. Miguelez, M. & Lewin. G. (2006). Efficacy of omega-3 fatty acids in preventing age-related macular degeneration. A systematic review. *Ophthalmology, 113*,1165-1173.

Hoegh-Guldberg, O. (1999). Climate change, coral bleaching, and the future of the world's coral reefs. *Marine Freshwater Research,50*(8), 839-866.

Holland, A. F., Sanger, D. M. Gawle, C. P. Lerberg, S. B. Santiago, M. S. Riekerk, G. H. M. Zimmerman, L. E. & Scott. G. I. (2004). Linkages between tidal creek ecosystems and the landscape and demographic features of their watersheds. *Journal of Experimental Marine Biology and Ecology, 298*(2),151-178.

IOOS: http://www.ocean.us/

Interstate Shellfish Sanitation Conference: http://www.issc.org/

Jackson, J. B. C., Kirby, M. X. Berger, W. H. Bjorndal, K. A. Botsford, L. W. Bourque, B. J. Bradbury, R. H. Cooke, R. Erlandson, J. Estes, J. A. Hughes, T. P. Kidwell, S. Lange, C. B. Lenihan, H. S. Pandolifi, J. M. Peterson, C. H. Steneck, R. S. Tegner, M. J. & Warner. R. R. (2001). Historical overfishing and the recent collapse of coastal ecosystems. *Science, 293*, 629-637.

Jansson, B., Asplund, L. & Olsson. M. (1987). Brominated flame retardants - Ubiquitous environmental pollutants? *Chemosphere, 16*, 2343-2349.

Jewett, E. B., Lopez, C. B., Dortch, Q. & Etheridge, S. M. (2007). National Assessment of Efforts to Predict and Respond to Harmful Algal Blooms in U.S. Waters. Interim Report. Interagency Working Group on Harmful Algal Blooms, *Hypoxia, and Human Health of the Joint Subcommittee on Ocean Science and Technology*. Washington, SC.

Joint Global Ocean Flux Study's data management system: http:// www1.whoi.edu/ jgdms_info.html

Joint Subcommittee on Ocean Science & Technology. (2007). Charting the course for ocean science in the United States: Research priorities for the next decade. 67 p. http://ocean.ceq. gov/about/sup_jsost_public_comment.html

Keim, M. E. (2006). Cyclones, tsunamis and human health: The key role of preparedness. *Oceanography, 19*(2), 40 – 49.

Keller, J. M., Kannan, K. Taniyasu, S. Yamashita, N. Day, R.D. Arendt, M.D. Segars, A.L. & Kucklick. J. R. (2005). *Perfluorinated compounds in the plasma of loggerhead and Kemp's ridley sea turtles from the southeastern coast of the U.S. Environmental Science and Technology, 39*, 9101-9108.

Kemp, W. M., Twilley, R. R. Stevenson, J. C. Boynton, W. R. & Means. J. C. (1983). The decline of submerged vascular plants in upper Chesapeake Bay: Summary of results concerning possible causes. *Marine Technology Society Journal, 17*, 78-89.

Key, P., Chung, K. Opatkiewicz, A. Wirth E. & Fulton. M. (2003). Toxicity of the insecticides Fipronil and Endosulfan to selected life stages of the grass shrimp (Palaemonetes pugio). *Bulletin of Environmental Contamination and Toxicology, 70*(3), 533-540.

Knap, A., Dewailly, E. Furgal, C. Galvin, J. Baden, D. Bowen, R.E. Depledge, M. Duguay, L. Fleming, L.E. Ford, T. Moser, F. Owen, R. Suk, W. & Unluata. U. (2002). Indicators of ocean health and human health: Developing a research framework. *Environmental Health Perspectives, 110*(9), 839- 845.

Kneib, R. T. (1997). The role of tidal marshes in the ecology of estuarine nekton. *Oceanography* and Marine Biology: An Annual Review, *35*, 163–220.

Kolpin, D. W., Furlong, E.T. Meyer, M.T. Thurmon, E.M. Zaugg, S.D. Barber, L.B. & Buxton. H. T. (2002). Pharmaceuticals, hormones, and other organic wastewater contaminants in U.S. streams, 1999-2000: A national reconnaissance. *Environmental Science and Technology, 36*(6), 1202-121 1.

Kovats, S., Patz, J. A. & Dobbins. D. (1998). Global climate change and environmental health: Proceedings of the 1997 annual conference of the Society for Occupational and Environmental Health. *International Journal of Occupational Environmental Health, 4*(1), 41-52.

Kuiken, T., Leighton, F. A. Fouchier, R. A. M. LeDuc, J. W. Peiris, J. S. M. Schudel, A. Stohr, K. & Osterhaus. A. D. M. E. (2005). Pathogen surveillance in animals. *Science, 309*,1680-1681.

Lafferty, K. D., Porter, J. W. & Ford. S. E. (2004). Are diseases increasing in the ocean? Annual Review of Ecology, *Evolution, and Systematics*, *35*, 31-54.

Laws, E. 2006. Case study: Cholera. *Oceanography, 19*(2),81-83.

Lee, S. H., Levy, D. A. Craun, G. F. Beach, M. J. & Calderon. R. L. (2002). Surveillance for waterborne-disease outbreaks: United States, 1999–2000. *Morbidity and Mortality Report, 51*(SS8),1- 28. http://www.cdc.gov/mmwr/PDF/ss/ss5108.pdf

Leeworthy, V.R. 2001. Preliminary estimates from Versions 1-6: Coastal recreation participation, national survey on recreation and the environment (NSRE) 2000. National Oceanic and Atmospheric Administration. NOAA Oceans and Coasts. *Special Projects Office*. 46 p. http://marineeconomics.noaa.gov/ NSRE/NSRE_V1-6_May.pdf

Leffler, J. W., Browdy, C. L. & Smith. T. I. J. (2006). Risk assessment--NOAA initiative examines fatty acids, contaminants in red drum, shrimp. Global Aquaculture Advocate, April/May issue:40-42.

Lerberg, S. B., Holland, A. F. & Sanger. D. M. (2000). Responses of tidal creek macrobenthic communities to the effects of watershed development. *Estuaries, 23*,838-853.

Lettieri, T. (2006). Recent applications of DNA microarray technology to toxicology and ecotoxicology. *Environmental Health Perspectives, 114*(1), 4-9.

Long, E., Scott, G. Kucklick, J. Fulton, M. Thompson, B. Carr, R. Biedenbach, J. Scott, K. Thursby, G. Chandler, G. Anderson, J. & Sloane. G. (1998). Magnitude and extent of sediment toxicity in selected estuaries of South Carolina and Georgia. National Ocean Service Technical Memorandum NOS ORCA 128. Silver Spring, MD. 289 p.

Luckenbach, M. W., Mann, R. & Wesson J. A. (eds.). (1999). Oyster reef habitat restoration. A synopsis and synthesis of approaches. Virginia Institute of Marine Science Press. Gloucester Point, VA. 358 p.

Mallin, M. A. (2006). *Wading in waste. Scientific American, 294*(6),53-59.

Mallin, M. A., Burkholder, J. M. Cahoon, L. B. & Posey. M. H. (2000). North and South Carolina coasts. *Marine Pollution Bulletin, 41*(1-6),56-75.

Marine cyanobacteria genome sequences: http://www.moore.org/ microgenome/ microb_list.asp

Marine eukaryotes genome sequencing: www.jgi.doe.gov Marine Genomics Project: www.marinegenomics.org

Mazet, J. A. K., Hunt, T. D. & Ziccardi. M. H. (2004). Assessment of the risk of zoonotic disease transmission to marine mammal workers and the public: Survey of occupational risks. Final report, Research Agreement No. K005486-01. Prepared for U.S. Marine Mammal Commission by the Wildlife Health Center, University of California, Davis. Unpublished report, 41 p + appendixes. http://www.sefsc.noaa. gov/PDFdocs/Marine_ Mammal_Zoonoses_Final_Report.pdf

McDonald, T. A. (2002). A perspective on the potential health risks of PBDEs. *Chemosphere, 46*, 745-755.

Miller, F., Thomalla, F. Downing, T. & Chadwick. M. (2006). Case study: Resilient ecosystems, healthy communities: Human health and sustainable ecosystems after the December 2004 tsunami. *Oceanography, 19*(2),50-51.

Mote Marine Laboratory real-time HAB reporting http://coolgate. mote.org/redtide/

Mozaffarian, D. & Rimm. E. B. (2006). Fish intake, contaminants, and human health: Evaluating the risks and the benefits. *Journal of the American Medical Association, 296*(15),1885-1899.

National Center for Education Statistics. (1996). Out-of-field teaching and educational equality. Statistical analysis report NCES 96-040. U.S. Department of Education, Office of Educational Research and Improvement Washington, DC. 70 p. http://www. teaching-point.net/Exhibit%20A/Ingersoll%20-%20Out%20of% 20Field%20Teaching.pdf

National Institute of Environmental Health Sciences & National Science Foundation. (2002). Centers for Oceans and Human Health request for applications (RFA ES-03-003). Research Triangle Park, NC. http://grants RFA-ES-03-003.html

National Marine Fisheries Service (NMFS). (2005). Fisheries of the United States, 2004. U.S. Department of Commerce. National Ocean and Atmospheric Administration, Silver Spring, MD. 109 p. http://www.st.nmfs.noaa.gov/st1/ fus/fus04/index.html

NMFS. (2002). *Annual report to congress on the status of U.S. fisheries -- 2001.* U.S. Department of Commerce. National Ocean and Atmospheric Administration, Silver Spring, MD. 142 pp.

National Oceanic & Atmospheric Administration (NOAA). (2006). Economic statistics for NOAA. National Oceanic and Atmospheric Administration. Program Planning and Integration Office of the NOAA Chief Economist. Silver Spring, MD. 68 p.

NOAA. (1999). Coastal hazardous waste site reviews. NOAA Office of Response and Restoration.127 p.

NOAA Coastwatch HAB news: coastwatch.noaa.gov/hab/bulletins_ ns.htm

NOAA's National Oceanographic Data Center: http://www.nodc. noaa.gov/

NOAA's Office of Marine and Aviation Operations: http://www. omao.noaa.gov/

National Research Council (NRC). (1999). From monsoons to microbes: Understanding the ocean's role in human health. *National Academies Press.* 2101 Constitution Avenue, N.W. Washington, DC 20418. 144 p.

NRC. (2001). Under the weather: Climate, ecosystems, and infectious disease. *National Academies Press.* 2101 Constitution Avenue, N.W. Washington, DC 20418. 160 p.

NRC. (2001). Grand challenges in environmental sciences. National Academies Press. 2101 Constitution Avenue, N.W. Washington, DC 20418. 96 p.

NRC. (2002). Marine biotechnology in the twenty-first century: Problems, promise, and products. *National Academies Press.* 2101 Constitution Avenue, N.W. Washington, DC 20418. 132 p.

NRC. (2003). NEON: Addressing the nation's environmental challenges. *National Academies Press.* 2101 Constitution Avenue, N.W. Washington, DC 20418. 108 p.

Natural Resources Defense Council (NRDC). (2006). Testing the waters 2006: Guide to water quality at vacation beaches. August 2006, Table 3, "U.S. Ocean, Bay, Great Lakes, and Some Freshwater Beach Closings/Advisories, 1994–2005". 349 p. http://www.nrdc.org/water/oceans

NRDC. (2005). Health risks and economic impacts of beach pollution. In Testing the Waters: A Guide to Water Quality at Vacation Beaches. Chapter 2. pp. 7-16.

National Response Plan: http://www.dhs.gov/xprepresp/committees/editorial_ 0566.shtm

Nesheim, M. C. & Yaktine Ann L. (eds.). (2006). Seafood choices: Balancing benefits and risks. Institute of Medicine of the National Academies. National Academies Press, 500 Fifth Street N.W. Lockbox 285. Washington, DC 20005. 608 p. Prepublication version available one-line. http://www.iom.edu/ CMS/3788/23788/37679.aspx

Newman, D. J. & Cragg. G. M. (2006.) Compounds from the ocean as drugs and drug leads. *Supplement to Chimica Oggi/Chemistry Today, 24*(5),42-47.

Newell, R. I. E. (2004). Ecosystem influences of natural and cultivated populations of suspension-feeding bivalve mollusks: A review. *Journal of Shellfish Research, 23*,51-61.

"Nowcasts" methodology for bacteria: http://pubs.water.usgs.gov/ tm6b5/

"Nowcast" for Huntington Beach, Ohio: http://www.ohionowcast. info/

O'Connor, T. & Ehler. C. (1991). Results from the NOAA Status and Trends Program on distribution and effects of chemical contamination in the coastal and estuarine United States. *Environmental Monitoring and Assessment, 17*,33-49.

Ocean. U. S.. (2006). Public health risks: Coastal observations for decision-making. Publication No. 15. Arlington, VA. 40 p.

Ocean.U.S. Web site: http://www.ocean.us

Ocean Carbon Biogeochemistry data management system http://ocb. whoi.edu/

Oken, E., Kleinman, K. P. Berland, W. E. Simon, S. R. Rich- Edwards, J. W. & Gillman. M. W. (2003). Decline in fish consumption among pregnant women after a national mercury advisory. *Obstetrics and Gynecology, 102*,346-351.

Olsen, G. W., Church, T. R. Miller, J. P. Burris, J. M. Hansen, K. J. Lundberg, J. K. Armitage, J. B. Herren, R. M. Medhdizadehkashi, Z. Nobiletti, J. B. O'Neill, E. M. Mandel, J. H. & Zobel. L. R. (2003a). Perfluorooctanesulfonate and other fluorochemicals in the serum of American Red Cross adult blood donors. *Environmental Health Perspectives, 111*,1892-1901.

ORION Executive Steering Committee. (2005). Observatories Initiative Science Plan. Washington, DC, 102 p.

Patz, J. A., Olson, S. H. & Gray. A. L. (2006). Climate change, oceans, and human health. *Oceanography, 19*(2),52-59.

Pew Oceans Commission. 2003. America's living oceans: Charting a course for sea change. A report to the nation. Pew Oceans Commission. Arlington, VA. 166 p. http://www.pewtrusts.org/ pdf/env_pew_oceans_final_report.pdf

Pine, J. C. (2006). Case study: Hurricane Katrina and oil spills: Impact on coastal and ocean environments. *Oceanography, 19*(2),37-39.

Porter, J. W. & Tougas. J. I. (2001). Reef ecosystems: Threats to their biodiversity. In S. Levin (ed.), *Encyclopedia of biodiversity, 5*,73–95. *Academic Press*, New York, NY.

Prospero, J. M. (2006). Case study: Saharan dust impacts and climate change. *Oceanography, 19*(2),60-61.

Red tide NGO: http://www.start1.com

Reeves, R. R. & Ragan. T. J. (2004). Executive Summary. In. Future directions in marine mammal research. A report of the Marine Mammal Commission consultation 4-7 Aug. 2003 Portland, OR. Marine Mammal Commission. Bethesda, MD. p. v-xiv. http://www.mmc.gov/reports/workshop/pdf/ Futuredirectionsreport.pdf

Rice, D. L., Dearry, A. & Garrison. D. L. (2004). Pioneering interdisciplinary research initiatives for oceans and human health. *EcoHealth, 1*,220-225.

Rose, D. P., & Connolly. J. M. (1999). Omega-3 fatty acids as cancer chemopreventive agents. *Pharmacology and Therapeutics, 83*(3),217-244.

Sanders, M., Sivertsen S. & Scott. G. (2002). Origin and distribution of polycyclic aromatic hydrocarbons in surficial sediments from the Savannah River. *Archives of Environmental Contamination and Toxicology, 43*,438-444.

Sandifer, P. A., Holland, A. F. Rowles, & Scott. G. I (2004). The oceans and human health. *Environmental Health Perspectives, 112*(8), A454-455.

Sanger, D. M., Holland, A. F. & Scott. G. I. (1999a). Tidal creek and salt marsh sediments in South Carolina coastal estuaries. I. Distribution of trace metals. *Archives of Environmental Contamination and Toxicology, 37*, 445-457.

Sanger, D. M., Holland, A. F. & Scott. G. I. (1999b). Tidal creek and salt marsh sediments in South Carolina coastal estuaries. II. Distribution of organic contaminants. *Archives of Environmental Contamination and Toxicology, 37*, 458-47 1.

Semantic Web: http://www.w3.org/2001/sw/ SensorNet: http://www.sensornet.gov/

Silvagni, P. A., Lowenstine, L. J. Spraker, T. Lipscomb, T. P. & Gulland. F. M. D. (2005). Pathology of domoic acid toxicity in California sea lions (Zalophus californianus). *Veterinary Pathology, 42*, 184-191.

Snape, J. R., Maund, S. J. Pickford, D. B. & Hutchinson. T. H. (2004). Ecotoxicogenomics: the challenge of integrating genomics into aquatic and terrestrial ecotoxicology. *Aquatic Toxicology, 67*(2),143-154.

Sogin, M. L., Morrison, H. G. Huber, J. A. Welch, D. M. Huse, S. M. Neal, P. R. Arrieta, J. M. & Herndl. G. J. (2006). Microbial diversity in the deep sea and the underexplored "rare biosphere". *Proceedings of the National Academies of Sciences, 103*(32),12115–12120.

Squires, S. (2006). Benefits of fish exceed risks, studies find. Experts advise 2 servings a week. Washington Post. Wednesday, October 18, 2006. Page A14. http://www.washingtonpost.com/wp-dyn/content/article/2006/10/17/AR200610 1700475. html

Stegeman, J. J., & Solow. A. R. (2002). Environmental health and the coastal zone. *Environmental Health Perspectives, 110*(11), A660-A661.

Taylor L. H., Latham, S. M. Woolhouse. M. E. J. (2001). Risk factors for human disease emergence. *Philosophical Transactions of the Royal Society B: Biological Sciences, 356*(1411), 983-989.

Steinback, S., Gentner, B. & Castle. J. (2004). The economic importance of marine angler expenditures in the United States. NOAA Professional Paper NMFS 2. 169 p.

Stumpf, R. P., Culver, M. E. Tester, P. A. Tomlinson, M. Kirkpatrick, G. J. Pederson, B. A. Truby, E. Ransibrahmanakul, V. & Soracco. M. (2003). Monitoring Karenia brevis blooms in the Gulf of Mexico using satellite ocean color imagery and other data. *Harmful Algae, 2*(2),147-160.

Thurman, H. V. (1994). Introductory *Oceanography*, seventh edition. Macmillan Publishing Company. New York, NY. 550 p.

Toledo, G., Green, W. Gonzalez, R. A. Christoffersen, L. Podar, M. Chang, H. W. Hemscheidt, T. Trapido-Rosenthal, H. G. Short, J. M. Bidigare, R. R. & Mathur. E. J. (2006). Case study: High throughput cultivation for isolation of novel marine microorganisms. *Oceanography, 19*(2), 120-125.

Turgeon, D. D., Asch, R. G. Causey, B. D. Dodge, R. E. Jaap, W. Banks, K. Delaney, J. Keller, B. D. Speiler, R. Matos, C. A. Garcia, J. R. Diaz, E. Catanzaro, D. Rogers, C. S. Hillis-Starr, Z. Nemeth, R. Taylor, M. Schmahl, G. P. Miller, M. W. Gulko, D. A. Maragos, J. E. Friedlander, A. M. Hunter, C. L. Brainard, R. S. Craig, P. Richond, R. H. Davis, G. Starmer, J. Trianni, M. Houk, P. Birkeland, C. E. Edward, A. Golbuu, Y. Gutierrez, J. Idechong, N. Paulay, G. Tafileichig, A. & Vander Velde. N. (2002). The state of coral reef ecosystems of the United States and Pacific Freely Associated States: 2002. National Oceanic and Atmospheric Administration, National Ocean Service, National Centers for Coastal Ocean Science. Silver Spring, MD. 265 p.

Tyson, F. L., Rice, D. L. & Dearry. A. (2004). Connecting the oceans and human health. *Environmental Health Perspectives, 112*(8), A455-456.

U.S. Commission on Ocean Policy. (2004). An ocean blueprint for the 21st century. Final report. Washington, DC. 676 p. http://www. oceancommission.gov/ documents/full_color

U.S. Ocean Action Plan. (2004). The Bush Administration's Response to the U.S. Commission on Ocean Policy. Washington, DC. December 17, 2004. 39 p.

USDA dietary site: http://www.health document/

U.S. Environmental Protection Agency (USEPA). (2002). National water quality inventory: 2000 report. Report # EPA-841-R-02- 001. USEPA Office of Water. Washington, DC. 214 p. http:// www.epa.gov/305b/2000report/

USEPA. (2001). National coastal condition report. EPA-620/R-01/005. U.S. EPA, Office of Research and Development/Office of Water. *Washington, DC.* 228 p. http://www.epa.gov/owow/oceans downloads.html

USEPA strategic planning: http://epa.gov/osp/myp.htm

U.S. Ocean Action Plan. (2004). The Bush Administration's Response to the U.S. Commission on Ocean Policy. *Washington, DC.* 39 pp.

Van Dolah, F.M. (2000). Marine algal toxins: origins, health effects, and their increased occurrence. *Environmental Health Perspectives, 108*(S1),133-141.

Venter Institute series of papers on the global ocean survey http://www.sorcerer2 expedition.org/version1/HTML/main.htm

Vitousek, P. M., Mooney, Jr., H A. Lubchenco, J. & Melillo. J.M. (1997). Human domination of Earth's ecosystems. *Science, 25*, 494-499.

Walker, N. D., Haag, A. Balasubramanian, S. Leben, R. Van Heereden, I. Kemp, P. & Mashriqui. H. (2006). Hurricane prediction: A century of advances. *Oceanography, 19*(2),24-36.

Wells, R. S., Rhinehart, H. L. Hansen, L. J. Sweeney, J. C. Townsend, F. I. Stone, R. Casper, D. R. Scott, M. D. Hohn, A. A. & Rowles. T. K. (2004). Bottlenose dolphins as marine ecosystem sentinels: Developing a health monitoring system. *EcoHealth, 1*(3), 246-254.

Willatts, P., & Forsyth. J. S. (2000). The role of long-chain polyunsaturated fatty acids in infant cognitive development. *Prostaglandins, Leukotrienes, and Essential Fatty Acids, 63*(2), 95-100.

Wilson, E. O. (1998). Consilience: the unity of knowledge. A.A Knopf. *New York*, NY. 352 p.

Wolf C. J., Fenton, S. E. Schmid, J. E. Calafat, A. M. Kuklenyik, Z. Bryant, X. A. Thibodeaux, J. Das, K. P. White, S. S. Lau, C. S. & Abbott. B. D. (2007). Developmental toxicity of perfluorooctanoic acid in the CD-1 mouse after cross-foster and restricted gestational exposures. *Toxicological Sciences*, *95*(2), 462-73.

Yang, Q., Abedi-Valugerdi, M. Xie, Y. Zhao, X. Y. Moller, G. Nelson, B. D. & DePierre. J. W. (2002). Potent suppression of the adaptive immune response in mice upon dietary exposure to the potent peroxisome proliferator, perfluorooctanoic acid. *International Immunopharmacology, 2*(2-3),389-97.

End Notes

[1] The term "ocean" refers to open ocean, coasts (including bays and estuaries), coastal watersheds, and Great Lakes.

[2] The term "seafood" refers to all consumer seafood products as well as recreational and subsistence take including marine mammal subsistence harvest by Alaska Natives.

[3] The term "drinking water" refers to fresh water supplies derived from the Great Lakes or from coastal seawater via desalinization.

[4] The term "ocean" refers to open ocean, coasts (including bays and estuaries), coastal watersheds, and Great Lakes.

[5] The term "seafood" refers to all consumer seafood products as well as recreational and subsistence take including marine mammal subsistence harvest by Alaska Natives.

[6] The term "drinking water" refers to fresh water supplies derived from the Great Lakes or from coastal seawater via desalinization.

CHAPTER SOURCES

The following chapters have been previously published:

Chapter 1 – This is an edited reformatted and augmented version of a NSTC Joint Subcommittee on ocean Science and Technology publication, dated January 26, 2007

Chapter 2 – This is an edited reformatted and augmented version of a The Interagency Working Group on Harmful Algal Blooms, Hypoxia, and Human Health publication, dated 2007.

INDEX

#

20th century, 134
21st century, 202

A

Abraham, 192, 195
academic settings, 167
access, 1, 7, 8, 28, 37, 54, 58, 69, 70, 105, 111, 119, 124, 137, 150, 160, 162, 163, 165, 166, 167, 177, 196
accountability, 55
accounting, 64, 83, 152
acid, 79, 83, 97, 98, 131, 132, 133, 141, 146, 148, 152, 193, 201, 202, 203
acidic, 26
acidity, 27
acidosis, 78
acoustics, 16
adaptation, 154, 182
additives, 73, 74
adulthood, 149
Advanced Reflectometer Surface Analysis, 185
advancement, 15, 69, 118, 120, 128, 132, 135, 145, 157, 164, 172, 175
advancements, 38, 60, 119, 135, 161
adverse effects, 76
advocacy, 50
aerosols, 33, 78, 80, 142, 175
aesthetic, vii, 67
aesthetics, 75
Africa, 112
agar, 122, 123
age, 59, 97, 130, 196
agencies, 6, 8, 10, 11, 39, 41, 47, 48, 49, 50, 51, 56, 57, 58, 59, 61, 67, 68, 69, 70, 85, 87, 90, 91, 93, 94, 97, 100, 101, 102, 105, 110, 112, 113, 115, 116, 117, 118, 124, 130, 135, 137, 140, 156, 157, 160, 165, 168, 172, 173, 174, 175, 176, 177, 178, 180, 181, 182, 183, 184, 185
agency actions, 5, 47
agency collaboration, 48
air emissions, 22
alanine, 96, 185
Alaska, 17, 74, 76, 83, 108, 203
Alaska Natives, 203
algae, 33, 41, 44, 60, 83, 94, 95, 97, 126, 127, 128, 134, 135, 158, 196
allergic reaction, 64, 79, 80, 81, 82
ALS, 96, 185
alternative energy, 15, 26
amino, 96, 116, 185, 194
amino acid, 96, 116, 185, 194
amino acids, 116
amnesia, 76
amniotic fluid, 146
amputation, 82
amyotrophic lateral sclerosis, 96
anemia, 79
animal disease, 143, 155, 178
animal diseases, 155
annual review, 56
anorexia, 77, 78
anthropology, 10
anti-asthma, 73
antibiotic, 109, 120, 135, 145
antibiotic resistance, 109, 135
anti-cancer, 33, 73, 136, 137
apoptosis, 148
appendicitis, 78
aquaculture, 13, 14, 15, 22, 73, 88, 102, 115, 124, 132, 133, 175
aquaria, 49, 168
aquatic systems, 120
aromatic hydrocarbons, 114
arthritis, 78

Artic Marine Assessment Program, 185
assessment, 5, 15, 16, 24, 30, 32, 34, 35, 37, 42, 45, 47, 53, 55, 56, 57, 103, 105, 111, 116, 162, 168, 175, 176, 177, 196, 198
assets, 4, 6, 33, 37, 38, 42, 48, 53, 166
assimilation, vii, 20, 32, 43, 45, 61, 67, 71
asthma, 73, 80, 113, 142, 195
atmosphere, 6, 20, 25, 26, 27, 28, 29, 30, 100, 157
atmospheric deposition, 34
authorities, 126
automation, 161
Autonomous Underwater Vehicle, 185
avian, 35, 89
avian influenza, 89
avoidance, 18
awareness, 24, 108
Azidothymidine (an antiretroviral drug), 135, 185
birds, 109, 117, 136, 143, 145, 146
birth weight, 76
bleaching, 26, 41, 196
blood, 77, 153, 200
blood pressure, 77
blueprint, 202
bone, 82, 135
bowel, 77
bradycardia, 80
brain, 73, 75, 130, 146, 148
brain damage, 75, 146
breakdown, 125
brevis, 83, 94, 116, 142, 151, 164, 201
brominated flame retardants, 125, 164
brucellosis, 109, 116
building blocks, 177
bun, 81

B

bacteria, 73, 92, 95, 103, 105, 106, 109, 111, 113, 115, 118, 119, 120, 121, 123, 124, 125, 126, 129, 134, 135, 137, 138, 145, 147, 150, 163, 164, 173, 200
bacterial infection, 81, 82, 153
bacterial pathogens, 92
bacterium, 107, 139
banks, 163, 165
barriers, 19, 39
base, 16, 127, 162
basic research, 98, 126, 157, 159, 184
behaviors, 97, 146
benefits, vii, 1, 2, 6, 7, 10, 13, 15, 29, 30, 32, 34, 64, 67, 69, 71, 73, 75, 89, 90, 91, 100, 102, 103, 108, 114, 124, 127, 129, 130, 132, 133, 154, 167, 174, 176, 178, 194, 199, 200
bioaccumulation, 34, 96, 155
bioactive materials, 163
biochemical processes, 35, 147
biodiversity, 3, 14, 31, 35, 43, 44, 71, 88, 114, 135, 137, 175, 181, 194, 200
bioinformatics, 36, 60, 98, 99, 109, 150, 161, 163, 177
biological responses, 151
biological samples, 164
biomarkers, 126
biomass, 16, 43
biomaterials, 182
biosafety, 135
biosensors, 45, 89
biosphere, 201
biosynthesis, 35
biotechnology, 97, 199

C

cables, 22
calibration, 37, 60
cancer, 32, 84, 135, 201
capacity building, 172
carbon, 25, 26, 27, 44, 45, 61
carbon dioxide, 26
cardiovascular disease, 73, 193
Caribbean, 17, 71, 73, 74
cartilage, 138
case study, 196
casting, 45
catastrophes, 172
catfish, 82
causal relationship, 140
CDC, 67, 68, 69, 83, 87, 91, 94, 97, 112, 113, 114, 118, 140, 141, 142, 143, 163, 168, 169, 172, 174, 176, 178, 179, 185, 187, 191
cell culture, 148
central nervous system, 109
cercaria, 80
certification, 55, 132
challenges, 6, 26, 34, 51, 56, 199
chemical, vii, 8, 16, 19, 27, 28, 30, 31, 35, 36, 37, 44, 60, 67, 69, 72, 73, 83, 84, 88, 96, 97, 99, 101, 103, 106, 108, 109, 114, 124, 126, 127, 130, 132, 133, 134, 135, 136, 140, 141, 143, 145, 149, 150, 153, 154, 155, 156, 157, 158, 161, 163, 165, 171, 173, 177, 179, 181, 182, 200
chemopreventive agents, 201
Chicago, 105
children, 34, 73, 80, 130, 131, 140
chlorophyll, 123, 160, 161
cholera, 34

chronic diseases, 109, 113, 185
chronic illness, 140
ciguatera, 95, 96, 134
ciguatera fish poisoning, 95
circulation, 4, 7, 14, 26, 27, 45, 46, 62, 84, 160, 163
cities, 26
citizens, 6, 39, 40, 173
City, 97
classes, 43, 126
climate, vii, 1, 3, 4, 6, 13, 14, 18, 23, 25, 26, 27, 28, 29, 30, 34, 38, 43, 44, 45, 46, 60, 61, 62, 67, 69, 71, 83, 101, 107, 117, 120, 132, 141, 154, 155, 156, 161, 175, 186, 193, 198, 200
clinical trials, 73, 135
closure, 158, 174
CO2, 26, 27, 28, 62
Coast Guard, 124
coastal communities, 5, 17, 19, 25, 75, 117, 140, 142, 156
coastal ecosystems, 13, 57, 75, 84, 89, 117, 144, 156, 197
coastal management, 59, 116
coastal region, 17, 25, 101, 106, 185
coastal watersheds, vii, 1, 5, 6, 7, 8, 13, 14, 27, 29, 33, 36, 63, 71, 124, 125, 203
coatings, 84, 124, 125
coding, 150
cognition, 34
cognitive development, 130, 202
collaboration, 5, 22, 47, 49, 51, 53, 54, 62, 69, 70, 91, 93, 94, 98, 103, 107, 108, 109, 112, 130, 137, 140, 142, 148, 150, 153, 164, 169, 175, 196
colleges, 184
color, 44, 60, 123, 157, 201, 202
coma, 79
combination therapy, 138
commerce, 21, 22, 23
commercial, 21, 22, 26, 35, 64, 73, 74, 83, 113, 127, 130, 137, 147, 151, 152, 153, 174
communication, 18, 20, 22, 28, 36, 39, 52, 71, 90, 110, 116, 132, 159, 167, 172, 178
communication strategies, 39
communities, 7, 10, 13, 14, 15, 19, 20, 33, 41, 50, 52, 54, 57, 58, 60, 62, 91, 99, 102, 105, 107, 111, 137, 141, 162, 180, 194, 198, 199
community, vii, 2, 5, 6, 8, 10, 19, 20, 34, 37, 38, 39, 40, 48, 52, 53, 54, 57, 58, 63, 68, 70, 71, 75, 85, 88, 91, 98, 100, 102, 103, 108, 110, 111, 112, 141, 142, 143, 153, 154, 158, 160, 161, 162, 166, 167, 170, 177, 178, 180, 183
community relations, 154
compatibility, 163
competition, 13, 86, 108, 109
competitive process, 102, 183
complement, 91, 108, 128, 151, 162, 172, 174, 180, 183
complex behaviors, 59
complex interactions, 1, 7
complexity, 28, 30, 68, 88, 178
compliance, 115, 118
composition, 59, 125, 170
compounds, 30, 34, 60, 68, 73, 74, 84, 92, 96, 97, 102, 103, 125, 126, 127, 135, 136, 137, 151, 154, 164, 197
comprehension, 9
computational capacity, 38
computer, 6, 105, 124, 150, 177, 184
computing, 161
conceptual model, 89
conductivity, 157
conference, 108, 198
configuration, 37
congress, 199
Congress, 56, 63, 68, 85, 86, 100, 128, 180, 182, 184
conjunctivitis, 116
connectivity, 48
consensus, 154
conservation, 6, 43, 64, 100, 114, 115, 183
constituents, 158, 172
Constitution, 199
consumers, 75, 90, 107, 114, 115, 116, 119, 126, 127, 132, 134, 140
consumption, 33, 34, 69, 71, 73, 76, 83, 90, 96, 102, 105, 107, 108, 113, 124, 125, 126, 130, 132, 133, 134, 141, 163, 175, 194, 195, 200
contaminant, 23, 30, 36, 41, 116, 117, 124, 132, 133
contaminated food, 77, 115
contaminated water, 78, 79, 83, 96, 149
contamination, 35, 84, 94, 95, 98, 102, 103, 111, 114, 115, 118, 119, 120, 122, 130, 132, 133, 153, 173, 200
Continental, 63
COOH, 184
cooking, 83
cooperation, 56, 110, 113, 184
coordination, 6, 23, 27, 36, 39, 44, 49, 50, 51, 52, 53, 55, 56, 70, 85, 96, 111, 150, 168, 178, 181, 182
coral reefs, 43, 196
cosmetic, 73
cosmetics, 125, 135
cost, 15, 18, 31, 114, 166, 196
cough, 77, 78, 80
covering, 156
crabs, 79, 141, 146
crops, 126
cross-fertilization, 91

cultivation, 135, 201
cultural heritage, 13
cultural practices, 97
culture, 35, 40, 96, 121, 135, 148, 165
current limit, 61
curricula, 39, 110, 170
curriculum, 168
curriculum development, 168
cycles, 8, 120, 132, 156
cycling, vii, 22, 27, 31, 34, 67, 71
cytometry, 95
cytotoxicity, 148

D

danger, 175
data collection, 23, 24, 28, 38, 68, 89, 159
data set, 16, 28, 37, 38, 42, 59, 61, 111, 143, 150, 157
data structure, 160
data transfer, 44
database, 28, 36, 155, 160, 162
death rate, 130
deaths, 83, 126
decision-making process, 53, 62
decomposition, 114
deficiency, 79
degradation, 17, 31, 71
dehydration, 78
dementia, 96
Department of Agriculture, 132, 189
Department of Commerce, 189
Department of Defense, 189
Department of Education, 199
Department of Energy, 190
Department of Health and Human Services, 113, 114, 185, 186, 190
Department of Homeland Security, 190
Department of Justice, 190
Department of the Interior, 12, 190
Department of Transportation, 190
deployments, 20
depression, 73
depth, 23, 92
dermatitis, 116
destruction, 13, 18
detectable, 146
detection, 15, 24, 33, 60, 106, 109, 114, 115, 116, 119, 120, 123, 124, 126, 132, 134, 142, 145, 148, 158, 161, 164, 165, 181
detection system, 15
developmental process, 148
diabetes, 73
diarrhea, 76, 77, 78, 79, 80, 109
diatoms, 98, 146
diet, 96, 130, 140, 193
dioxin, 195
dioxin-like compounds, 195
directors, 103, 112
disaster, 171, 172
discharges, 118, 124, 125, 163
discomfort, 78, 79
diseases, 34, 35, 68, 84, 87, 89, 91, 94, 107, 109, 113, 114, 115, 116, 118, 134, 140, 141, 145, 154, 155, 167, 171, 172, 175, 180, 181, 194, 198
dispersion, 22
displacement, 44
distress, 80
distribution, 2, 14, 26, 44, 60, 76, 92, 126, 128, 136, 141, 155, 157, 170, 175, 200, 201
diversity, 5, 13, 30, 31, 33, 37, 40, 92, 121, 126, 135, 136, 144, 182, 201
dizziness, 79
DNA, 41, 92, 98, 104, 107, 119, 120, 136, 148, 151, 153, 165, 172, 186, 198
DOC, 189
dogs, 196
DOI, 59, 62, 191
Domestic Policy Council, 190
dominance, 73
donors, 200
dosage, 150
draft, 160
drinking water, vii, 33, 67, 75, 105, 114, 118, 122, 129, 178, 203
drug discovery, 67, 175
drug therapy, 115
drugs, 33, 73, 84, 89, 114, 115, 134, 135, 136, 137, 200
drying, 74

E

early warning, 69, 75, 89, 100, 101, 117, 130, 147, 150, 174, 176
earthquakes, 117
ecology, 35, 44, 59, 92, 99, 106, 115, 116, 127, 151, 160, 184, 194, 197
economic activity, 73
economic development, 5, 50
economic engine, vii, 1, 71
economic losses, 90, 126
economic resources, 15
economic status, 154
economics, 10, 32, 40

ecosystem, vii, 1, 2, 4, 5, 6, 7, 8, 9, 13, 14, 16, 17, 18, 19, 22, 24, 26, 27, 30, 31, 32, 34, 35, 41, 42, 43, 44, 50, 51, 57, 58, 59, 60, 63, 64, 67, 69, 84, 85, 100, 101, 104, 105, 108, 120, 121, 124, 125, 126, 129, 132, 143, 145, 147, 149, 150, 155, 160, 176, 192, 194, 202
education, 18, 32, 39, 40, 51, 53, 54, 55, 69, 70, 92, 101, 102, 105, 119, 142, 159, 167, 168, 175, 177, 178, 182, 185
educational institutions, 55
educational opportunities, 99, 107, 183
educators, 39, 53, 54, 168, 170
effluents, 121
ELISA, 164, 165, 186
emergency, 54, 70, 83, 168, 171, 172, 178
emergency response, 168, 172
Emerging Chemicals of Concern, 186
employees, 183
employment, 89
employment opportunities, 89
endocrine, 34, 76, 125, 181
end-users, 102
energy, 7, 12, 13, 14, 15, 16, 22, 23, 25, 29, 31, 43, 117, 157
Energy Policy Act of 2005, 63
energy supply, 12
engineering, 20, 40, 55, 60, 63
environment, vii, 1, 2, 4, 5, 6, 7, 8, 9, 21, 22, 23, 24, 34, 38, 41, 44, 83, 88, 92, 94, 101, 113, 114, 115, 117, 125, 126, 127, 128, 143, 144, 152, 194, 195, 198
environmental change, 14, 23, 37, 59
environmental conditions, 22, 23, 24, 57, 104, 124, 144, 151
environmental degradation, 120
environmental factors, 2, 23, 92, 97, 175
environmental impact, 3, 13, 22, 23, 25, 115, 173
environmental protection, 22, 24, 114
Environmental Protection Agency (EPA), 58, 62, 68, 69, 83, 84, 87, 91, 97, 105, 111, 112, 113, 114, 117, 118, 122, 124, 131, 142, 156, 163, 168, 172, 173, 174, 176, 178, 179, 184, 186, 188, 192, 195, 202
environmental quality, 104, 114
environmental stress, 60, 100, 101, 103, 143
environmental threats, 114, 144
environmental variables, 24
enzyme, 164
enzyme-linked immunosorbent assay, 164
Enzyme-Linked Immunosorbent Assay, 186
enzymes, 35, 73, 152
eosinophilia, 79
epidemiologic studies, 35, 91, 140, 141, 142, 145
epidemiology, 69, 99, 120, 140, 195
equality, 199
equipment, 8, 103, 167, 182
erosion, 18
estuarine systems, 54
EU, 196
Europe, 139
European Commission, 92
European Union, 62
evaporation, 26
evidence, 14, 25, 38, 78, 120, 130
evolution, 18, 20, 35, 92
Exclusive Economic Zone (EEZ), 13, 71
expenditures, 201
experimental design, 140
expertise, 50, 52, 68, 88, 91, 104, 105, 112, 137, 172, 178, 183
exposure, 34, 68, 72, 83, 90, 96, 97, 105, 113, 118, 124, 140, 141, 142, 143, 144, 146, 147, 148, 149, 150, 153, 154, 163, 169, 175, 193, 203
expressed sequence tag, 152
extraction, 13, 14, 15, 22, 23, 29, 30, 115, 165, 172

F

fainting, 82
farms, 76
fatty acids, 73, 103, 129, 132, 133, 193, 196, 198, 201
FDA, 68, 69, 76, 87, 91, 107, 113, 114, 115, 130, 131, 139, 168, 170, 172, 173, 174, 176, 178, 179, 185, 186, 187, 188
feces, 77, 78, 79
federal agency, 51, 113, 116, 128, 184
federal funds, 136
Federal Geographic Data Committee, 160, 186
federal government, 6, 22, 57, 97, 107
federal regulations, 113
feed additives, 33
FEMA, 172
fermentation, 182
fever, 77, 78, 79, 109
field trials, 104
financial, 106, 183
financial support, 106, 183
fire retardants, 103
fish, 1, 7, 13, 14, 68, 72, 73, 75, 77, 79, 80, 82, 83, 107, 115, 117, 124, 125, 126, 130, 131, 132, 133, 134, 135, 136, 145, 149, 158, 173, 181, 193, 194, 200, 201
Fish and Wildlife Service, 84, 189
fisheries, 6, 14, 15, 26, 59, 73, 83, 99, 127, 130, 152, 199

fishing, 13, 14, 21, 24, 35, 43, 76, 126
flame, 34, 84, 124, 150, 194, 195, 197
flame retardants, 34, 84, 124, 150, 194, 195, 197
flexibility, 56
flooding, 17, 18, 25, 27
Florida Department of Health, 169, 186
fluctuations, 61
fluid, 44, 60, 78, 122
fluorescence, 161
focus groups, 52
food, vii, 1, 5, 13, 14, 29, 33, 34, 77, 78, 83, 84, 95, 96, 102, 113, 114, 115, 117, 118, 123, 124, 127, 134, 135, 143, 144, 146, 155, 192, 193
food additives, 135
food chain, 83, 84, 96, 127, 144
food poisoning, 77
food production, 14
food products, 102
food safety, 115
food web, 34, 95, 124, 127, 134, 143, 146, 155, 193
Ford, 194, 197, 198
forecasting, vii, 1, 2, 4, 7, 8, 14, 23, 30, 37, 41, 42, 43, 44, 45, 61, 88, 105, 106, 110, 112, 132, 157, 163
formal education, 39, 54, 99
formation, 1, 7, 18, 27, 114, 142
fouling, 126, 127
foundations, 47
fragments, 148
France, 153
freedom, 8, 21
freshwater, 13, 26, 27, 28, 83, 96, 113, 114, 121, 125, 152, 185, 193
fruits, 77
funding, 8, 17, 54, 56, 85, 93, 97, 103, 111, 112, 169, 174, 181, 183
funds, 100, 108, 109, 116
fungi, 118, 122, 134, 135

G

gastroenteritis, 123, 196
gene expression, 148, 151, 153, 164
genes, 95, 107, 115, 147, 151, 152, 153
genetic diversity, 92
genetic factors, 119
genetic marker, 92, 107
genetic screening, 44
genetics, 106
genome, 107, 152, 198
genomics, 8, 35, 36, 44, 60, 69, 70, 88, 92, 94, 100, 103, 147, 149, 150, 151, 160, 177, 181, 183, 201
genus, 92
geographic information systems (GIS), 16
geography, 40
geology, 8, 40
Georgia, 110, 186, 198
Germany, 194
global climate change, 120, 155
Global Earth Observing System of Systems (GEOSS), 37
global economy, 21
global ocean model configurations, 38
global scale, 10, 16, 19, 159
goods and services, 29, 31, 58, 101
governance, 5, 6, 13, 30, 48, 49, 50, 52
governments, 7, 105, 168
GPC, 138
grades, 170
graduate students, 91, 99
grants, 69, 85, 86, 100, 101, 102, 103, 108, 110, 112, 147, 174, 181, 183, 199
grass, 126, 127, 197
grasses, 137
grassroots, 91
gravity, 160
Great Lakes, vii, 1, 2, 5, 6, 7, 8, 13, 14, 15, 21, 22, 23, 29, 33, 34, 35, 37, 49, 51, 63, 75, 83, 100, 101, 105, 106, 108, 110, 111, 113, 114, 117, 124, 129, 142, 157, 163, 182, 186, 199, 203
green alga, 80
greenhouse, 26
greenhouse gases, 26
gross domestic product, 71
groundwater, 122
Group on Earth Observations, 51, 117, 158, 186
growth, 125, 126, 127, 136, 149
growth rate, 149
guidance, vii, 1, 38, 55, 56, 194
guidelines, 47, 55, 160
Gulf Coast, 17, 120
Gulf of Mexico, 51, 71, 108, 114, 116, 142, 173, 201

H

habitat, 2, 13, 14, 71, 192, 194, 198
habitat quality, 14
habitats, 13, 26, 28, 32, 35, 69, 75, 89, 101, 104, 117, 119, 135, 143, 144, 145, 146, 147, 151, 176, 194
harbors, 21, 22, 23
harmful algal blooms (HABs), vii, 32, 67, 75, 184, 192
harmful effects, 103, 126, 130, 133
harvesting, 76, 113, 140, 141, 152, 174
Hawaii, 17, 94, 96, 137, 138, 188, 196
hazardous waste, 199

hazards, 1, 2, 5, 7, 16, 17, 18, 19, 20, 25, 32, 33, 34, 49, 57, 113, 115, 124, 126, 153, 172, 173, 181, 185
headache, 77, 78, 79, 82
health care, 71, 90, 168, 175, 178
health care costs, 71, 90
health condition, 113
health effects, 34, 69, 84, 91, 94, 95, 113, 129, 142, 143, 144, 145, 146, 150, 160, 175, 202
health information, 117, 169
health practitioners, 141
health problems, 87
health promotion, 102
health risks, 33, 34, 41, 69, 89, 90, 97, 103, 105, 106, 129, 132, 145, 155, 156, 167, 169, 175, 179, 184, 198, 200
health status, 104, 142, 150, 153
heart disease, 124, 130, 149
heart rate, 82
heavy metals, 75, 129
height, 16, 45
hemisphere, 62
hemorrhage, 79
HHS, 113, 185, 186
high school, 170
High Throughput Culturing, 186
highlands, 147
HIV, 135, 186
hives, 80
homeland defense, 21
homeland security, 111, 156, 161
homes, 167, 173
hormones, 197
host, 111, 135, 170, 178
human activity, 29, 121
human behavior, 18, 154
human dimensions, 185, 193
human exposure, 69, 84, 92, 96, 97, 113, 140, 152, 153, 154, 175
human health, 3, 4, 32, 33, 34, 35, 36, 44, 49, 67, 68, 69, 70, 73, 75, 85, 86, 87, 88, 89, 90, 91, 92, 93, 94, 95, 100, 101, 102, 103, 104, 105, 106, 107, 109, 110, 111, 112, 113, 114, 116, 117, 120, 121, 124, 129, 132, 133, 140, 143, 144, 145, 146, 149, 152, 153, 154, 155, 156, 158, 160, 162, 163, 165, 167, 168, 172, 173, 174, 177, 178, 179, 180, 181, 182, 183, 184, 192, 193, 194, 195, 196, 197, 199, 200, 201, 202
Hungary, 196
Hunter, 193, 202
Hurricane Katrina, 17, 42, 91, 114, 142, 172, 173, 200
hurricanes, vii, 16, 17, 26, 30, 34, 41, 49, 64, 67, 140, 167, 172
hydrocarbons, 34
hypotension, 80
hypothesis, 94, 159
hypoxia, 41, 100, 181, 183

I

ideal, 39, 55
identification, 19, 41, 54, 58, 100, 107, 112, 119, 128, 134, 145, 148, 165, 177, 178, 182, 183
identity, 75
image, 161
imagery, 201
imagination, 9
immune function, 84
immune response, 76, 203
immune system, 35, 125, 140
immunocompromised, 123
impact assessment, 45
Impact Assessment, 64
implants, 135
imported products, 73
imports, 132
improvements, 19, 23, 28, 44, 60, 90, 141
incidence, 34, 76, 83, 96
income, 15, 23
indigenous peoples, 140
indirect effect, 107, 141
individuals, 8, 32, 34, 40, 63, 109, 140, 183
industries, 184
industry, 10, 13, 22, 48, 52, 53, 54, 57, 72, 115, 132, 135, 156, 183
infection, 77, 79, 116, 154
inflammation, 135
inflammatory disease, 73
information technology, 16, 32, 40, 105, 162
infrastructure, 4, 5, 8, 16, 17, 18, 19, 20, 23, 24, 26, 31, 32, 35, 36, 47, 48, 50, 53, 54, 55, 57, 60, 68, 69, 88, 105, 111, 150, 157, 159, 162, 165, 174, 175, 176, 177, 196
ingestion, 77, 78, 79, 96, 123
inhibitor, 139
initiation, 86, 93, 151, 158
injuries, 113, 185
injury, 116, 140
institutions, 5, 48, 49, 53, 71, 86, 91, 100, 102, 107, 108, 110, 181
integration, 6, 15, 20, 24, 32, 35, 42, 57, 62, 63, 68, 70, 89, 120, 145, 150, 157, 162, 170, 175, 177, 179
integrity, 147

intellectual capital, 55
intellectual property, 137
interagency coordination, 70, 179
interface, 29, 30, 32, 34, 38, 51, 165, 192
International Atomic Energy Agency, 117, 186
international standards, 24
international trade, 114
interoperability, 24, 37, 57, 160
intervention, 90
invertebrates, 145
investment, 8, 16, 32, 47, 92, 114
investments, 7, 30, 48, 68, 73
Ireland, 139
iron, 27
islands, 19, 94
isolation, 120, 134, 201
issues, 2, 4, 6, 9, 10, 11, 14, 21, 22, 36, 38, 49, 50, 51, 53, 54, 70, 91, 94, 100, 108, 111, 112, 113, 114, 116, 117, 140, 141, 145, 155, 157, 163, 168, 171, 196

J

Japan, 92
jaundice, 78
joint pain, 109
Joint Subcommittee on Ocean Science and Technology, v, 1, 40, 68, 87, 187, 189, 197
jurisdiction, 174

K

kidney, 75, 145

L

La Niña phenomena, 7
laboratory studies, 151
lakes, 15, 126
landscape, 18, 57, 197
landscapes, 4, 13, 41
languages, 169
larvae, 149
lead, 8, 20, 25, 27, 32, 41, 61, 90, 99, 100, 114, 115, 118, 125, 145, 146, 150, 154, 172, 178
leadership, 10, 39, 47, 109, 114
Leahy, 190
leaks, 172
learning, 40, 55, 109, 130, 146
learning environment, 40
legislation, 182
lesions, 146
leukemia, 135
levees, 19
lice, 80
lifelong learning, 40
light, 4, 44
Lion, 80
liquid chromatography, 127
literacy, 39, 50, 70, 71, 90, 102, 110, 167, 179
liver, 76, 80, 125
liver damage, 80
local authorities, 171
local community, 108
local government, 10, 11, 117
Louisiana, 173
lung cancer, 138
lying, 25

M

macular degeneration, 73, 196
magazines, 40
magnitude, 19, 28, 173
majority, 22, 47, 107, 115, 124, 132, 158, 170
malaria, 112, 185
mammal, 64, 106, 115, 116, 146, 163, 171, 195, 198, 200, 203
mammals, 75, 84, 98, 100, 107, 109, 115, 116, 117, 125, 126, 135, 143, 144, 145, 146, 158, 175, 181, 183, 193
man, 24, 71
management, vii, 1, 2, 3, 4, 5, 6, 7, 13, 14, 15, 19, 20, 23, 25, 26, 27, 28, 30, 31, 32, 35, 37, 38, 39, 40, 41, 42, 43, 49, 51, 52, 53, 54, 55, 58, 59, 60, 63, 64, 69, 83, 89, 100, 101, 103, 107, 108, 111, 112, 119, 126, 140, 150, 155, 156, 157, 158, 159, 160, 162, 163, 165, 168, 174, 177, 196, 197, 200
manipulation, 32
man-made disasters, 71
mapping, 14, 16, 21, 24, 32, 57, 58, 59, 92, 151
marine environment, vii, 1, 5, 7, 8, 9, 22, 29, 49, 67, 71, 73, 92, 96, 100, 110, 121, 125, 126, 137, 145, 149, 153, 165, 181, 183, 195
marine fish, 125, 133
Marine Mammal Commission, 68, 115, 187, 191, 198, 200
Marine operations, 21
marine species, 35, 69, 75, 89, 104, 127, 135, 143, 151, 176
marketplace, 73, 115
marsh, 201
Maryland, 110
mass, 25, 27, 134, 135, 136

materials, 19, 34, 35, 39, 44, 70, 73, 98, 105, 118, 121, 125, 133, 135, 136, 144, 164, 165, 177, 179
mathematics, 55, 63, 184
matter, 71
McGillicuddy, 33, 192
measurement, 55, 129, 153, 163
measurements, 8, 14, 37, 44, 60, 61, 93, 116, 157, 158, 159, 160, 163, 164, 165, 177
meat, 74, 123
media, 54, 102, 167, 168, 179
medical, 36, 73, 88, 91, 94, 134, 141, 168, 181, 183
medical science, 88, 181
medication, 73
medicine, 39, 73, 88, 94, 100, 140, 181, 183
Mediterranean, 32, 138
melanoma, 138
melting, 25, 27, 28
memory, 79
memory loss, 79
meningitis, 78
mercury, 34, 75, 124, 130, 131, 133, 200
Mercury, 193
messages, 173
metabolism, 84, 138, 188
metabolites, 135, 136, 164
metals, 124, 201
methodology, 124, 200
Mexico, 114, 194
Miami, 94, 142, 151, 166, 169, 188
mice, 125, 148, 152, 203
microarray technology, 198
microbial communities, 125
microorganisms, vii, 67, 75, 104, 106, 115, 134, 135, 158, 201
microscopy, 98
midbrain, 148
mid-career, 70, 178
migration, 22
military, 15, 21, 22, 23
mineral resources, 14, 117
mission, 48, 57, 92, 94, 100, 101, 108, 113, 114, 116, 165, 172, 176, 178, 184
missions, 36, 37, 47, 49, 56, 91, 156, 165, 183
misuse, 29
model system, 92
models, 2, 3, 4, 14, 15, 18, 19, 20, 23, 24, 27, 28, 30, 31, 34, 35, 37, 38, 39, 40, 41, 42, 43, 46, 53, 57, 58, 59, 60, 61, 69, 88, 89, 92, 93, 102, 106, 110, 111, 119, 120, 122, 126, 129, 132, 141, 142, 144, 145, 148, 149, 150, 154, 156, 157, 162, 163, 175, 176, 181, 196
Moderate Resolution Imaging Spectroradiometer, 160
moderates, vii, 67
MODIS, 160, 161, 187
moisture, 26
mole, 146
molecular biology, 106, 135
mollusks, 73, 95, 97, 135, 143, 145, 200
momentum, 26
Moon, 80
morbidity, 149
mortality, 116, 146, 149
mucoid, 77, 78
museums, 40, 168
mussels, 79, 131, 143

N

nanotechnology, 8
National Aeronautics and Space Administration, 59, 68, 116, 187, 191
national borders, 49
National Center for Education Statistics, 199
National Institute for Occupational Safety and Health, 113, 185
National Institute of Standards and Technology, 104, 163, 187
National Institutes of Health, 187
National Research Council, 8, 63, 85, 108, 181, 187, 199
National Response Plan, 171, 200
National Seafood Inspection Program, 76, 115
national security, vii, 1, 21
Native Americans, 64
natural disaster, 4, 16, 17, 57, 88, 117
natural disasters, 4, 16, 57, 88, 117
natural hazards, 2, 16, 17, 18, 19, 23, 34, 49, 69, 156, 175
natural resources, 2, 6, 15
natural science, 10, 16, 40
natural sciences, 10, 16, 40
nausea, 77, 78, 79, 80, 82
nematode, 79
nervous system, 145
Netherlands, 196
networking, 157, 160
New England, 172, 174
next generation, 157
nitrogen, 27, 164
Norway, 153
nuclear magnetic resonance, 135
nursing, 130, 131
nutrient, vii, 30, 41, 64, 67, 71, 95, 96, 163
nutrients, 33, 35, 44, 60, 73, 105, 129, 130, 157

O

occupational risks, 198
Ocean Action Plan (OAP), 47
ocean circulation, 4, 26, 27, 45, 160
Ocean circulation, 7
oceans, 26, 67, 68, 74, 86, 87, 90, 91, 94, 100, 102, 104, 108, 109, 110, 135, 140, 141, 152, 153, 155, 160, 167, 170, 179, 180, 181, 182, 183, 184, 193, 194, 195, 199, 200, 201, 202
Office of Management and Budget, 56, 62, 190
officials, 4, 41, 51, 105, 106, 111, 120, 148, 154, 162, 168, 172, 179
oil, 13, 22, 58, 124, 132, 133, 149, 172, 179, 200
oil spill, 22, 124, 149, 172, 200
omega-3, 73, 129, 132, 133, 196
operations, 2, 3, 15, 16, 19, 21, 22, 23, 24, 27, 35, 42, 101, 120, 156, 185
opportunities, vii, 9, 13, 26, 35, 39, 49, 50, 55, 56, 57, 67, 69, 70, 102, 103, 108, 110, 118, 120, 126, 128, 132, 135, 144, 145, 150, 154, 155, 156, 157, 162, 164, 165, 168, 172, 175, 178, 182, 192
optical properties, 60
optimization, 165
organ, 110, 119
organism, 94, 129, 144, 147, 149, 151, 153, 156
ornamental plants, 126
otitis media, 78
outreach, 40, 52, 54, 56, 57, 68, 69, 70, 92, 98, 99, 101, 103, 105, 108, 110, 142, 166, 167, 168, 174, 175, 177, 178, 185
oversight, 156
oxidative stress, 148, 152
oxygen, vii, 67
oyster, 76, 123, 153, 194
oysters, 77, 79, 107, 123, 131, 143, 144, 153

P

Pacific, 7, 17, 26, 27, 46, 63, 71, 79, 94, 97, 99, 108, 116, 117, 130, 139, 141, 142, 153, 188, 195, 202
pain, 73, 77, 78, 79, 80, 81, 82, 135, 139, 145
paradigm shift, 9
parallel, 8, 61, 127
paralysis, 76, 79, 80
parameter estimates, 59
parasites, 103, 109
paresthesias, 80
participants, 53, 57, 70, 87, 111, 179, 183
pathogenesis, 120, 132, 144, 146
pathogens, 33, 34, 35, 41, 44, 60, 69, 75, 76, 83, 88, 89, 92, 95, 100, 101, 103, 104, 106, 107, 108, 109, 113, 115, 117, 118, 119, 120, 121, 122, 123, 129, 130, 132, 135, 140, 141, 142, 143, 144, 145, 147, 150, 154, 155, 156, 157, 158, 164, 165, 173, 175, 183
pathology, 150, 168
pathways, 34, 94, 118, 126, 146, 152
PCBs, 75, 125
PCR, 129
peer review, 48, 91
penicillin, 134, 135
permission, 74, 83, 128, 136, 144
permit, 8, 21, 24, 133, 153
pests, 126
pharmaceutical, 135, 164
pharmaceuticals, vii, 13, 14, 15, 33, 35, 67, 69, 72, 73, 74, 89, 91, 94, 96, 102, 103, 112, 114, 115, 124, 125, 127, 136, 150, 152, 164, 175, 181, 195
pharmacology, 35
pharyngitis, 78
physical characteristics, 41
physical environment, 16
physical properties, 27
physics, 8, 40
physiology, 84, 106, 144, 150, 151, 160
phytoplankton, 44, 192
Plan, 157, 186, 194
plankton, 45, 60, 106
plants, 133, 134, 197
platform, 167
poison, 80, 98
polar, 27, 29, 36, 71
policy, 4, 6, 9, 19, 20, 25, 28, 40, 41, 49, 54, 71, 89, 111, 154
policy options, 154
policy responses, 19
pollutants, 24, 27, 35, 75, 100, 101, 105, 124, 129, 130, 149, 158, 163, 183, 197
pollution, 29, 67, 71, 75, 101, 104, 111, 114, 115, 119, 121, 124, 143, 149, 152, 165, 194, 199
Polybrominated Diethyl Ether, 188
polychlorinated biphenyl, 124
polychlorinated biphenyls (PCBs), 124
polycyclic aromatic hydrocarbon, 124, 149, 201
polyunsaturated fat, 202
polyunsaturated fatty acids, 202
ponds, 126
population, 17, 18, 22, 26, 40, 59, 64, 71, 84, 92, 95, 104, 116, 140, 146, 158, 160, 193, 194, 195
population growth, 71, 84
portfolio, 6, 8
Portugal, 85
positive interactions, 70
potential benefits, 3, 34, 88

power generation, 15
precipitation, 155, 157
precursor cells, 148
predation, 30
predictability, 10, 28
preparation, 79, 83, 115, 168
preparedness, 197
President, 56, 87, 179, 180, 182, 190
prevention, 23, 32, 42, 71, 89, 90, 113, 116, 124, 143, 169, 177
Prince William Sound, 76
principles, 47
probe, 150
producers, 94, 132
professional development, 168
professional-development, 39
professionals, 39, 70, 104, 110, 154, 178
profit, 173, 181, 183
project, 3, 5, 28, 47, 92, 94, 95, 96, 97, 99, 116, 152, 193
proliferation, 120, 148, 155
propagation, 18
prosperity, 21, 184
protected areas, 14, 44
protection, vii, 6, 21, 58, 60, 67, 71, 113, 115, 130, 140, 148, 158
proteins, 147
proteomics, 36, 70, 88, 147, 150, 177, 181
prototype, 101
prototypes, 44
Pseudo-nitzschia, 97, 98, 131
public awareness, 167, 169, 171, 179
public health, 4, 6, 10, 17, 18, 19, 27, 36, 40, 41, 60, 68, 70, 71, 83, 87, 89, 92, 93, 94, 97, 100, 101, 102, 103, 104, 105, 106, 107, 108, 110, 111, 112, 113, 116, 120, 126, 127, 129, 132, 140, 141, 142, 147, 148, 150, 152, 154, 156, 158, 160, 162, 163, 167, 168, 169, 172, 174, 177, 178, 179, 180, 181, 185, 192, 195, 196
public interest, 9
public policy, 47, 105
public safety, 42
public service, 154
public-private partnerships, 44, 52
Puerto Rico, 127
purification, 134

Q

quality assurance, 163, 164
quality of life, 5, 6, 73, 117
quantification, 41, 58, 114, 120, 124, 128, 145, 165, 177
quotas, 14

R

radar, 157
rainfall, 106, 122, 157
rash, 78, 79, 80, 116
reactions, 151
reagents, 35, 136
real time, 36, 93, 174
recognition, 85
recommendations, 6, 56, 91, 113, 115, 127, 130, 133, 158, 162, 169, 172, 181, 182, 184
recovery, 16, 18, 30, 31, 42
recreation, vii, 1, 5, 13, 14, 33, 35, 83, 152, 198
recreational, vii, 13, 35, 64, 67, 73, 74, 83, 94, 105, 106, 111, 113, 118, 120, 123, 129, 130, 141, 142, 152, 153, 154, 163, 170, 174, 178, 203
regions of the world, 25
regulations, 115
regulatory agencies, 105, 165, 172
relevance, 51, 114, 162
remote sensing, 60, 99, 117, 157, 159, 169
replication, 14
reproduction, 75
requirements, 19, 21, 23, 40, 54, 87, 108, 160, 161, 162
researchers, 8, 22, 38, 39, 49, 52, 60, 70, 83, 89, 91, 92, 93, 94, 95, 96, 97, 99, 102, 104, 108, 110, 111, 119, 124, 127, 130, 133, 134, 147, 148, 149, 150, 151, 152, 153, 157, 158, 164, 165, 166, 167, 168, 171, 172, 173, 174, 177, 178, 179
reserves, 13
residues, 115
resilience, 4, 5, 17, 18, 19, 31, 41, 57, 145
resistance, 18, 19
resolution, 14, 16, 28, 38, 58, 157, 158
resource management, 6, 15, 16, 38, 40, 44, 50, 52, 54, 55, 120, 154
resource policies, 15
resources, 1, 3, 5, 6, 7, 13, 14, 15, 16, 20, 32, 33, 34, 35, 36, 47, 48, 49, 50, 54, 64, 73, 84, 88, 89, 101, 102, 110, 114, 116, 135, 136, 152, 156, 161, 162, 168, 172, 174, 176, 179, 181, 182
respiration, 34
respiratory problems, 113
response, 2, 4, 17, 18, 19, 20, 38, 41, 57, 58, 59, 60, 69, 70, 86, 99, 101, 102, 104, 108, 113, 118, 119, 124, 141, 148, 149, 150, 151, 160, 167, 169, 171, 172, 173, 178, 193, 196
restaurants, 132
restoration, 13, 19, 24, 31, 114, 198
restructuring, 19

revenue, 72, 152
rheumatoid arthritis, 73
rhythm, 149
rights, 22, 31
rings, 46
risk, 2, 3, 4, 5, 8, 10, 16, 17, 18, 19, 23, 34, 36, 68, 72, 73, 89, 95, 97, 100, 101, 102, 103, 105, 107, 108, 112, 113, 114, 115, 116, 120, 121, 126, 128, 129, 130, 131, 132, 133, 135, 140, 147, 148, 154, 156, 158, 162, 163, 168, 172, 174, 194, 195, 198, 199, 200, 201
risk-taking, 8
robotics, 8
roots, 136
routes, 22, 23, 27, 96
ROVs, 156, 165, 177
Royal Society, 64, 201
RTS, 176
rubella, 78
runoff, 34, 75, 124, 125, 143, 147, 157, 158, 163

S

safety, 4, 19, 23, 41, 44, 67, 69, 88, 100, 107, 114, 115, 119, 132, 156, 163, 178, 185
salinity, 27, 28, 157, 195
salmon, 73, 79, 195
saltwater, 82
SARS, 188
saturated fat, 130
scale system, 157
school, 39, 167, 179
science, 1, 6, 7, 8, 9, 10, 17, 24, 31, 39, 40, 47, 48, 50, 51, 52, 54, 55, 57, 62, 63, 91, 98, 112, 115, 116, 154, 158, 160, 167, 175, 178, 184, 196, 197
scientific knowledge, 5, 8
scientific publications, 101
scientific understanding, 31, 58, 68, 87, 180
scope, 1, 6, 9, 21, 55, 100, 108, 140
seafood, 13, 32, 33, 34, 64, 67, 68, 69, 71, 72, 73, 74, 75, 77, 80, 90, 100, 102, 103, 104, 107, 108, 111, 113, 114, 115, 119, 123, 124, 125, 126, 129, 130, 132, 133, 134, 140, 152, 154, 156, 166, 172, 175, 178, 181, 183, 192, 203
sea-level, 17, 18, 23, 25, 27, 28, 61, 117
Secretary of Commerce, 182, 183, 184
security, 5, 21, 24, 62
sediment, 18, 23, 28, 30, 34, 41, 64, 84, 120, 121, 144, 177, 198
sediments, 69, 120, 124, 126, 135, 137, 157, 165, 172, 173, 175, 201
seizure, 79
selenium, 133
seminars, 110
sensations, 79
sensing, 16, 19, 36, 37, 60, 62, 156, 159
sensitivity, 46, 150
sensors, 8, 16, 23, 24, 28, 34, 35, 36, 39, 44, 60, 70, 88, 103, 119, 120, 128, 132, 147, 156, 157, 158, 159, 160, 165, 175, 177, 182
sepsis, 77, 79
septic tank, 122
sequencing, 8, 36, 92, 107, 150, 151, 152, 198
serum, 200
services, 29, 31, 50, 58, 71, 73, 92, 101, 108, 132, 185
Severe Acute Respiratory Syndrome, 188
sewage, 75, 77, 78, 95, 118, 121, 122, 124, 125, 179, 194
shellfish, 33, 35, 75, 79, 80, 83, 92, 93, 98, 103, 107, 114, 115, 118, 119, 131, 141, 153, 158, 159, 163, 170, 173, 174, 192
shoreline, vii, 18, 19, 22, 29, 31, 35, 67, 71
shrimp, 127, 133, 197, 198
signs, 78
simulation, 6, 7
skin, 34, 73, 74, 77, 81, 82, 116
social activities, 152
social sciences, 10, 16, 19, 32, 35, 39, 55, 184
social services, 16
social theory, 37
society, vii, 1, 2, 4, 6, 7, 8, 9, 13, 14, 16, 21, 25, 26, 28, 30, 31, 49, 100, 101
sociology, 10, 32
software, 28, 38
Solutions to Avoid Red Tide, 169, 188
South America, 62
specialists, 166
species, 2, 11, 13, 14, 15, 22, 23, 24, 30, 31, 35, 36, 41, 43, 59, 60, 64, 69, 73, 77, 78, 80, 82, 84, 93, 94, 95, 97, 98, 101, 103, 104, 107, 108, 109, 112, 115, 119, 121, 125, 126, 128, 130, 131, 134, 135, 136, 139, 143, 144, 145, 146, 148, 149, 150, 151, 158, 164, 168, 170, 171, 188, 196
specter, 126
spine, 81, 82
spleen, 77
sponge, 81, 135, 137, 138, 139
Spring, 105, 194, 198, 199, 202
St. Petersburg, 192
stability, 2, 14, 15, 18, 23, 30, 115
stakeholder groups, 105
stakeholders, 22, 37, 49, 51, 54, 56, 99, 111, 158
standardization, 164
state, 1, 5, 6, 7, 10, 11, 22, 24, 25, 28, 36, 38, 45, 47, 48, 49, 51, 54, 57, 58, 59, 61, 62, 63, 64, 89, 91,

93, 94, 97, 101, 110, 112, 113, 117, 156, 166, 170, 171, 172, 173, 174, 177, 202
states, 28, 43, 49, 51, 56, 67, 68, 111, 113, 131
statistics, 99, 199
Stewardship of Natural and Cultural Ocean Resources, 2, 10, 13, 36, 42
storage, 27, 114, 150, 165
storms, 17, 18, 26, 117
stormwater, 34, 118
strategic planning, 202
stress, 31, 103, 152
stressors, 30, 41, 71, 107, 108, 121, 145, 149, 150
stroke, 130
structure, 19, 30, 31, 50, 51, 52, 53, 55, 59, 96, 196
subgroups, 50
subsistence, 17, 35, 64, 74, 75, 90, 102, 116, 130, 132, 152, 154, 203
supplier, 71
suppression, 203
surveillance, 36, 41, 42, 64, 69, 83, 89, 110, 111, 112, 113, 114, 120, 121, 123, 129, 140, 141, 143, 144, 145, 156, 158, 169, 175, 178, 198
survival, 125, 146, 149
susceptibility, 96, 97, 154
sustainability, 2, 5, 14, 15, 29, 31, 58, 132
sustainable energy, 15
Sweden, 194
swelling, 82
Switzerland, 192
symptoms, 83, 118
synthesis, 43, 103, 134, 138, 147, 157, 182, 198

T

tapeworm, 79
target, 34, 103, 126, 148, 154, 175
taxa, 194
taxonomy, 32
teachers, 39, 105, 168, 170
teams, 52, 104, 173
technical support, 63, 106
techniques, 8, 9, 14, 32, 38, 41, 43, 89, 93, 96, 97, 99, 103, 106, 120, 121, 129, 132, 150, 152, 157, 165, 172, 174, 182
technological advances, 21, 48, 157
technological developments, 15
technologies, 2, 15, 16, 19, 23, 36, 39, 44, 60, 101, 102, 103, 104, 107, 110, 118, 132, 133, 150, 157, 161, 181, 182
technology, 1, 5, 7, 8, 9, 16, 17, 24, 32, 39, 48, 50, 53, 54, 55, 57, 63, 68, 70, 89, 93, 102, 104, 105, 107, 112, 132, 137, 142, 147, 148, 161, 168, 177, 178
technology transfer, 68, 89, 178
teeth, 82
temperature, 27, 30, 32, 45, 114, 120, 123, 154, 155, 163
terrestrial ecosystems, 29
terrorism, 172
testing, 100, 135, 158, 175, 176, 183
therapeutics, 135, 136
threats, 3, 5, 17, 33, 35, 41, 69, 71, 84, 88, 89, 90, 93, 94, 96, 100, 101, 103, 107, 109, 111, 113, 118, 128, 140, 141, 143, 144, 145, 147, 151, 154, 156, 157, 158, 161, 164, 165, 169, 170, 173, 176, 177, 178, 181, 185
tides, 83, 93, 113, 142, 169, 195
time frame, 185
time periods, 24, 137
time series, 155, 196
tissue, 32, 115, 116, 126, 131, 163, 177
tobacco, 149
tobacco smoke, 149
tourism, 13, 14, 35, 72, 75, 83, 152
toxic shock syndrome, 79
toxicity, 97, 98, 114, 125, 126, 127, 132, 149, 150, 170, 174, 195, 198, 201, 202
toxicology, 35, 69, 99, 140, 176, 198
toxin, 33, 36, 91, 92, 94, 96, 97, 98, 113, 115, 126, 131, 135, 146, 151, 152, 165
trace elements, 27
trade, 15, 22
trade-off, 15
trainees, 85, 182
training, 16, 38, 39, 40, 54, 55, 68, 70, 89, 99, 101, 105, 110, 111, 112, 168, 177, 178, 179, 181, 183
training programs, 55
traits, 92, 154
translation, 5, 18, 20, 38, 39, 47, 53, 54, 55, 98
transmission, 67, 77, 78, 107, 198
transport, 18, 23, 41, 44, 60, 71, 94, 97, 105, 120, 121, 124, 157, 158, 163, 165
transportation, 3, 13, 14, 21, 22, 23, 64, 114
treaties, 52
treatment, 23, 32, 48, 122, 141, 168, 177
trial, 73
tuberculosis, 116
tumor, 76, 125, 135, 136

U

U.S. Army Corps of Engineers, 114
U.S. Commission on Ocean Policy (USCOP), 5
U.S. Department of Commerce, 132, 199
U.S. Department of the Interior, 63, 117, 132
U.S. economy, vii, 21, 67

U.S. Geological Survey, 64, 68, 117, 189
underlying mechanisms, 97
underserved students, 168
underwater vehicles, 156
UNESCO, 117, 189, 192, 196
uniform, 164
United, vii, 1, 2, 7, 10, 13, 17, 22, 24, 48, 49, 57, 58, 62, 64, 68, 71, 87, 89, 101, 106, 108, 112, 117, 124, 146, 185, 186, 189, 192, 193, 195, 196, 197, 198, 199, 200, 201, 202
United Kingdom, 62
United Nations, 117, 186, 189
United States, vii, 1, 2, 7, 10, 13, 17, 22, 24, 48, 49, 57, 58, 64, 68, 71, 87, 89, 101, 106, 108, 112, 118, 124, 146, 185, 189, 192, 193, 195, 196, 197, 198, 199, 200, 201, 202
universities, 49, 55, 94, 97, 105, 113, 142, 170, 184
upper respiratory tract, 122
urban, 84, 85, 118, 193
urbanization, 195
urine, 77
USDA, 91, 92, 107, 113, 189, 202

V

validation, 14, 18, 37, 60, 163, 164, 177
valuation, 3, 15, 31, 35, 88, 154
vapor, 164
variables, 27, 28, 37, 59, 120, 155, 195
variations, 4, 25, 45, 59, 64, 97, 160
vector, 45, 91, 94, 181
vegetables, 77, 79
vehicles, 8, 16, 23, 31, 125, 156
velocity, 44, 60
venue, 49, 52
vertebrates, 97, 145
vessels, 23, 24, 31, 37, 127, 137, 156, 165, 166, 167, 177
viral pathogens, 115
virtual organization, 195
viruses, 78, 95, 103, 104, 106, 109, 115, 118, 129, 150, 165, 173
viscera, 82
vision, 5, 7, 47, 68, 91, 100, 177
visualization, 6
vitamin B1, 79
vomiting, 77, 78, 79, 80
vulnerability, 2, 17, 18, 19, 20, 30, 41, 42, 57, 58, 154

W

walking, 83
warning systems, 35
Washington, 63, 64, 65, 97, 99, 106, 108, 130, 131, 141, 148, 173, 188, 189, 196, 197, 199, 200, 201, 202
waste, vii, 29, 67, 71, 198
waste disposal, 29
wastewater, 13, 34, 75, 104, 118, 122, 197
water, 13, 14, 20, 22, 25, 26, 27, 30, 31, 34, 37, 41, 42, 44, 58, 71, 73, 75, 77, 78, 79, 80, 83, 84, 91, 92, 93, 94, 95, 97, 101, 104, 105, 106, 111, 112, 113, 114, 117, 118, 119, 120, 121, 122, 123, 124, 129, 131, 138, 140, 142, 143, 144, 149, 157, 158, 163, 165, 166, 169, 170, 172, 173, 177, 181, 192, 194, 195, 199, 200, 202, 203
watershed, 14, 15, 19, 28, 30, 31, 34, 37, 49, 51, 58, 147, 198
waterways, 21, 22, 23, 64
weakness, 78, 79, 80, 109
wealth, 29, 33, 87, 144, 170
wear, 142
weather patterns, 25, 45
web, 143, 146, 196
weight loss, 79
welfare, 48, 147, 154, 184
well-being, 3, 33, 35, 71, 75, 88, 175
wells, 122
wetlands, 18, 19, 26, 29, 31
whales, 109
white blood cell count, 77
wild animals, 104
wildlife, 34, 35, 75, 89, 120, 121, 125, 132, 143, 164, 194
Wisconsin, 17
workers, 89, 116, 132, 140, 198
workforce, 8, 16, 32, 39, 40, 53, 54, 55, 167
working groups, 53
workplace, 113, 185
World Health Organization (WHO), 117, 189, 192
worldwide, 109, 115, 117
worms, 73

X

xanthophyll, 33

Y

yield, 92, 135, 160